Sustainability and Interdisciplinarity

This book series is inspired by the Springer SDGs Programme which aims to "connect the researchers who are tackling the world's toughest challenges with the practitioners in policy and business who desperately need those insights to achieve their goals in improving the world, by making our publishing activities more visible to our key communities through a variety of channels.

Xin Liu • Fang Zhong • Jun Zhang

Design for Sustainability

Springer

Xin Liu
Academy of Arts and Design
Tsinghua University
Beijing, China

Fang Zhong
School of Design
Sichuan Fine Arts Institute
Chongqing, China

Jun Zhang
School of Design
Hunan University
Changsha, Hunan, China

ISSN 3005-1312 ISSN 3005-1320 (electronic)
Sustainability and Interdisciplinarity
ISBN 978-981-96-8656-8 ISBN 978-981-96-8657-5 (eBook)
https://doi.org/10.1007/978-981-96-8657-5

Jointly published with Tsinghua University Press
The print edition is not for sale in China (Mainland). Customers from China (Mainland) please order the print book from: Tsinghua University Press.
The original submitted manuscript has been translated into English. The translation was done using artificial intelligence. A subsequent revision was performed by the author(s) to further refine the work and to ensure that the translation is appropriate concerning content and scientific correctness. It may, however, read stylistically different from a conventional translation.
Translation from the Chinese language edition: "可持续设计" by Xin Liu et al., © Tsinghua University Press 2022. Published by Tsinghua University Press. All Rights Reserved.

This Springer imprint is published by the registered company Springer Nature Singapore Pte Ltd.
The registered company address is: 152 Beach Road, #21-01/04 Gateway East, Singapore 189721, Singapore

Thank you to Xia Nan, Wang Bosong, Song Jiajia, and Maurizio Vrenna for their contributions to this book.

Dedicated to those who aspire to change the world: designers, engineers, educators, entrepreneurs, and dreamers.

Foreword

Sustainability is no longer a peripheral concern but a defining principle for humanity's future. It is a call to action that transcends disciplines, geographies, and social and economic contexts, urging us to rethink the way we produce, consume, and design. Among the various fields contributing to this global shift, design occupies a pivotal role. It bridges creativity and implementation feasibility, offering solutions that address environmental challenges, foster social equity and inclusion, and promote economic prosperity. As designers' researchers, educators, and practitioners, we stand at a critical juncture where our collective creativity can either perpetuate unsustainable practices or forge pathways toward a resilient and sustainable future for all.

The publication of *Design for Sustainability* is a landmark moment in this journey. Authored by Professors Liu Xin, Jun Zhang, and Zhong Fang, this book represents a significant contribution to the field of Design for Sustainability (DfS). Originally written in Chinese, it has already established itself as a key resource for researchers, educators, and students in China. The release of its English edition marks a crucial step in amplifying its reach and impact, enabling global audiences to benefit from its insights and approaches. As someone deeply invested in the promotion of design for sustainability in research, education and practice, I am honored to write this foreword.

The book's structure is both comprehensive and forward-thinking, addressing the historical context, theoretical underpinnings, and practical approaches and strategies of DfS. It navigates the evolution of sustainability from a conceptual framework to a practical imperative, illustrating how design can act as a catalyst for system changes. Through its twelve chapters, the book explores key concepts such as product Life Cycle Design, circular economy principles, biomimicry, Product-Service Systems (PSS), social innovation, and systemic thinking. These topics are not merely theoretical; they are actionable frameworks that guide designers in tackling complex, multidimensional challenges.

One of the book's most compelling features is its contextualization within China's unique sociocultural and economic landscape. By incorporating localized case studies, the authors demonstrate how global sustainability principles can be adapted to address specific regional challenges and opportunities. For example, the book highlights innovative practices emerging in China, offering valuable lessons for other regions. These case studies not only enrich the book's narrative but also underscore the importance of cultural and institutional factors in shaping sustainable design solutions.

The inclusion of these localized perspectives aligns seamlessly with the ethos of the Learning Network for Sustainability (LeNS), a network counting on more than 150 universities worldwide, aiming at diffusing DfS knowledge-base and know-how, through a learning-by-sharing, multipolar, and open-access ethos. As the founder of LeNS, I have had the privilege of working with a diverse community of scholars and practitioners, including the authors of this book. Their contributions to the LeNS network have been instrumental in fostering a deeper understanding of how sustainability can be integrated into design research, education, and practice. This book stands as a testament to the power of collaboration, mutual exchange, and co-creation, embodying the core values that LeNS seeks to promote.

A key strength of the book lies in its emphasis on systemic approaches to sustainability. The authors recognize that addressing today's sustainability challenges requires system innovations that account for the interconnectedness of environmental, social, and economic dimensions. By advocating for integrative design strategies, the book equips readers with the tools needed to navigate this complexity. It encourages the design community to move beyond isolated interventions and to embrace a systemic mindset, where each decision contributes to a broader vision of sustainability.

Another notable aspect of the book is its forward-looking perspective. While it provides a thorough analysis of current theories and practices, it also anticipates emerging trends and challenges. The chapters on Product-Service Systems (PSS) and Distributed Economies (DE) design for sustainability, for example, explore how those models can drive sustainability by win-win solutions, i.e. environmentally, socio-ethically, and economically beneficial. By addressing these forward-thinking topics, the book positions itself as a valuable resource not only for today's readers but also for future generations of designers, educators, and researchers.

The publication of this book in English is particularly significant in fostering cross-cultural dialogue and knowledge exchange. It enables readers from diverse backgrounds to engage with China's unique contributions to DfS while also contributing their own perspectives to the global discourse. This exchange of ideas is essential for advancing sustainability, as it allows us to learn from one another's successes and challenges, building a collective understanding of what it means to design sustainability for all.

Beyond its scholarly contributions, the book carries a profound message of hope and empowerment. It reminds us that design can be a transformative force that shapes the way we live, work, and interact with the world. By embracing sustainability as a guiding principle, designers can become agents of change, driving the transitions needed to address the urgent challenges of our time. This message resonates deeply with the ethos of LeNS and with my own belief in the power of design to create positive environmental and social impacts, with economic benefits.

As this book reaches an international audience, its potential to inspire and educate is boundless. It serves as a vital resource for students, educators, researchers, and professionals across disciplines, equipping them with the knowledge-base and know-how needed to contribute to the transition towards a sustainable society. More importantly, it fosters a sense of responsibility and agency, encouraging readers to see themselves as active participants in the journey toward sustainability.

In closing, I extend my heartfelt congratulations to Professors Liu Xin, Jun Zhang, and Zhong Fang for their outstanding work. Their dedication to advancing design for sustainability in education is evident on every page of this book, and their contributions will undoubtedly leave a lasting impact on the field. I am grateful for the opportunity to contribute this foreword and to support the dissemination of their invaluable insights to a global audience.

May this book serve as a source of inspiration, guidance, and collaboration for all who read it. Together, let us harness the power of design to envision and contribute to an environmentally, socially, and economically sustainable society for all.

Design for Sustainability, Design Department, Politecnico di Milano Learning Network for Sustainability (LeNS) I LeNSlab Polimi
Milan, Italy

Carlo Vezzoli

Foreword

1. The transition towards sustainability is a social learning process in which we all, that is, the whole humanity, have much to experience and learn.

Today, almost 40 years after we began this journey, we know that the social systems we must generate must be, first of all, *ecological* and *just*. Where "ecological" means rich in diversity and mutual relationships between humans and between humans and non-human beings (and this at every scale: from that of individuals to those of communities and societies as a whole). And where "just" means that in them all human beings have the same rights and that everyone must be given the same opportunities (which also means that, given the inequalities found today, it is necessary to offer more possibilities to those who, up to now, have had less).

Ecological and just are two adjectives whose philosophical meaning is profound and open to a variety of interpretations. In indicating them as the main characteristics of the transition that we would like to see implemented, it does not mean proposing a model to be reproduced, but a path to undertake. Everyone starts from where they are. Each with its own specificity. Knowing that we all need to learn from everyone else. And, at the same time, we all have something to bring to others.

This book seems to me to do so: it is open to all the ideas that, over time and in all regions of the world, have emerged regarding sustainability and design. And, at the same time, it introduces themes that are, or could become, the specific contribution that China can bring to this great social conversation on the future of the Planet.

By doing so, it seems to me that it proposes design for sustainability as an ecosystem of ideas that coexist, supporting or contrasting each other (because they refer to complementary aspects of the same reality or because they refer to different approaches and visions of the world). A set of proposals constitutes the current, vast, and articulated array of visions and tools available to guide action for the future of our planet.

All this, in my opinion, makes it an important and useful book. A book that is, at the same time, a compendium of everything that has been said, written, and experienced up to now on the topic of design for sustainability, and an indicator of the themes that represent its future.

2. In his Preface to this same book, Liu Guanzhong writes: *"The standard for evaluating the innovativeness of a rational design comes from ancient Chinese philosophy and entails assessing what is appropriate, moderate, and suitable, rather than the infinite pursuit of desire and optimal solutions"*. I believe that this statement is of great relevance, in China and everywhere. And I believe that, when faced with the challenges of the future, it should be carefully considered.

Of course, what "appropriate, moderate, and suitable" means today must be discussed and new shared visions of the qualities we want to seek must be found. But many years of work in the field of sustainability clearly tell us that it is necessary to radically change the ways of thinking and doing that are dominated today. Design for sustainability can and must follow all the different strategies indicated in this book, but at the basis there must be the search for a renewed wisdom that has to lead us to redefine the criteria with which we judge the quality of life.

In my opinion, based on many different experiences, this redefinition must be made on the basis of three keywords: *collaboration, proximity,* and *care*. Where "collaboration" means that only by working together we can hope to achieve ecological and just sustainability. "Proximity" means that sustainability is to be sought starting from its molecular physical dimension, creating rich and diversified proximity systems: proximity systems habitable by us humans, in harmony with the non-human entities that constitute the network of life on our Planet. "Care" means attention, empathy, and closeness and refers to the quality of relationships between human beings and between human beings and their environment. Care relationships which, in order to exist, require giving oneself the necessary time to build them.

All this may seem to be shared by everyone (who could declare themselves in opposition to the need to collaborate, be close and take care?). But if these keywords are taken seriously, with all their practical implications, they appear in stark contrast to the lines of action currently dominant in modernized societies. In fact, the economic and cultural models on which they are based and the technologies they produce push in a diametrically opposite direction, towards societies of isolated individuals competing with each other, towards desertified proximity systems, and towards careless interactions.

In opposition to these dominated trends, design for sustainability is therefore design for collaboration, proximity, and care. More precisely, it is the design of proximity systems that make collaborative and caring relationships possible and probable, which, as has been said, does not constitute a complete model to be replicated. But it indicates a direction, leaving it open, case by case, to decide how one should actually operate, combining ideas and practices that come from the past with new sensitivities and new technological possibilities. It's not an easy path, but I think this book helps you get started.

Design for Social Innovation a ELISAVA
Barcelona School of Design and Engineering
Barcelona, Spain

Ezio Manzini

Foreword I: China's Solution for Sustainable Design

German philosopher Martin Heidegger once expounded a view of history stating that the development of history does not follow a linear model of past–present–future, but rather is interactive, whereby based on the future, we examine the present and reflect on the past. When we approach history in this way, the past, present, and future generate an increased number of interactions and collisions.

Industrial Design Has Always Responded to the Needs of the Times

During the historical period spanning the end of the 19th century through the beginning of the 20th century, Chinese society underwent tremendous transformations. Around 1876, as part of the Qing government's New Policy, Western mechanical drawing was introduced to China. Over the following 50 years, the Chinese made breakthroughs in developing light industrial products, establishing new industries, and transplanting Western industrial design thinking. Relying on the industrial policies and regulations of that time, as well as the nation's propensity for a market of nationally produced goods, China established some of its own industries.

After the founding of The People's Republic of China, the thinking and practices of industrial design did not keep pace with the international community, but were nevertheless not completely isolated. In 1949, Chinese designers learned the Soviet model. In the 1960s, China advocated for constructing a self-reliant industrial system. In the late 1970s, the government selectively introduced advanced international production equipment and technology. Then the country underwent reforms, opening up and implementing a socialist market economy. Recently, practitioners have advocated for the use of industrial design to enhance the core competitiveness of enterprises and achieve innovation-driven transformation and development. Chinese industrial design approaches and practices have always responded to the needs of the times, albeit in different forms, playing a huge role, with fiery creativity, in each generation.

In the 1950s and 1960s, a large number of industrial products that determined the fate of the nation were developed, such as the Liberation truck, China's first 10,000-ton hydraulic press. These products are miracles created by enthusiasm, courage, wisdom, and countless attempts by all participants under circumstances in which reference materials are extremely scarce.

In the mid-1970s, Chinese industrial manufacturing enterprises completed a technical equipment upgrade and transformation to varying degrees to meet the demand for improving product quality. Simultaneously, they developed technology to tackle key problems, overcoming a large number of difficulties in product manufacturing. Their designers also discovered that the lack of change in products over the course of many years meant that they had significantly fallen behind the living requirements of the population.

In the 1980s, during the period of reforms and opening up, China implemented a policy of exchanging market for technology and Chinese products expanded their production capacity, generating profits and ensuring the expansion of new production. Enterprise directors gradually realized the importance of industrial design and sought to change the situation whereby engineers and technicians only guest-starred in industrial design, thus increasing investment in industrial design. This period marks an era of the arrival of new household appliances represented by washing machines, refrigerators, and air conditioners. Although these products are still developed and designed based on the copying similar foreign products, they nonetheless liberated people from their traditional lifestyles and significantly improved the quality of life.

Since the beginning of the 21st century, industrial design has begun to focus on upgrading and branding, and China's independent industrial design has begun to develop rapidly.

Chinese Industrial Design's Purpose Is Not Only to Reflect, but Also to Discover

History recedes. Old technologies and products are inevitably replaced by new ones, but design culture can settle and undergo redevelopment. Examining Chinese industrial design as part of the long history of global industrial design development, our focus is on the new content of Chinese design culture, not on traditional achievements, and more importantly, on discovering the dawn of future Chinese industrial design. The purpose of studying Chinese industrial design is: "not only to reflect, but also to discover; not only for nostalgia, but also to look forward to surpassing [the past]."[1] Almost everyone in the Chinese industrial design field knows that design has two purposes. One is to design products for people. The other is to generate economic benefits. Currently, however, an increasing number of scholars question whether these two purposes constitute the entire mission of industrial design. If there is a lack of noble spiritual guidance, "people orientation" may cause people's desires to expand indefinitely. If we emphasize enhancing economic benefits too much, it causes ruthless damage to the environment. Therefore, shaping the spirit of the times through industrial design and understanding sustainable development is where the current spirit of design lies.

European countries have inherited and promoted the brilliant culture of Ancient Greece, and the European Renaissance illuminated the world with culture. After the Industrial Revolution, technology caught up. Europe has dominated industrial design in the 20th century. This dominance includes British designer William Morris's Arts and Crafts Movement against the production relationship of the Industrial Revolution, the structural mechanism of the German Manufacturing Alliance's industrial system, the Bauhaus Movement's advocacy for integrating technology and art, and the baptism of the modern design movement. Industrial design is everywhere under the sky, all in European style. Today, European scholars claim in writing that China does not have industrial design.

Contributing Chinese Industrial Design Culture to the World

In 1918, German scholar Oswald Arnold Gottfried Spengler published *The Decline of the West*. Later, British historian Arnold Joseph Toynbee opposed Eurocentrism. The latter, through dialogue with Japanese social activist Daisaku Ikeda, introduced a discussion of Eastern culture as an interaction between Eastern and Western cultures. Objectively speaking, the practices of Europe, America, Japan, and South Korea all have aspects worthy of learning. However, we must recognize that only by seriously studying and cataloguing China's industrial design practices and approaches can we contribute to our industrial design culture while recognizing the industrial design thinking and achievements of various countries in the world.

In fact, most of the designs passed down in China have been symbolic. Too many designs concern ancient Chinese palaces and imperial power. Why are most of these designs for emperors and generals? Because the best ones were created for them and then preserved. Most of the things created by the common people have been lost and people do not pay attention to folk objects. The former belongings of emperors and generals are dazzling and stimulate the eye. These things can satisfy people's desires, but do not necessarily to meet people's needs. Why do we think that folk things are simple? Because they are used to solve problems.

Over thousands of years, humankind has created a brilliant civilization. Both the tools of ancient times, like stone axes and the spaceships that humans use to travel in space today were created by humans to adapt to the environment and transform nature, creating today and designing tomorrow. From humankind's most basic motivation—survival and food—and the planned exploration of the mysteries of the universe, entering the era of artificial intelligence, design always reflects humankind's understanding of the world, that is, the objectification of humans' essential power. The concept of humankind understanding the world and transforming nature is the culmination of development from low to high, from simple to complex, from single to overlapping, from coherent to network, and also the result of continuous creation.

However, without the dominance of concepts, there would be no distinction between humans and animals, no creation, and there could have been no emergence of human civilization. One characteristic of the Marxism is linking the understanding of nature with labor practices, because the labor process objectifies humans' essential power. The labor practice Marx mentions is the design concept and creation process of humankind.

Design Is the Reflection of a Way of Life and the Spiritual World

As early as 1985, I wrote that "design is the design of the way of life," which not only refers to material life, but also reflects the spiritual world. Design in the industrial age had to reflect the characteristics and traditions of this period, which were the result of human technology and culture to date, and contradict natural laws and humans' natural attributes. These results and contradictions bring to design new ways of living. Design provides the energy and motivation to create the future.

What we call tradition is created, not inherited. As designers, we must recognize our historical mission to create. If we only indulge in the past and passively inherit tradition, history will eliminate us. Tradition is relative to the present and the future. If we do not focus on the creation of the contemporary and what is to come, there will be no continuation and development of tradition. History will also be interrupted. Studying the current practical problems in China, solving the problem of common prosperity for a population of 1.4 billion people, no longer just a few rich people as in the past, but improving the welfare of most, requires a lifestyle that is contemporary, socialist, Chinese, which will also be the style of China in the 21st and 22nd centuries. Our descendants will call this tradition.

Culture emerges from past ways of life—how geography, climate, production, economy, politics, customs, and values influence people's lives. Their existence determines their consciousness (including culture). Therefore, the most critical contributions to culture are region or space in certain time periods. In short, culture has a spatial attribute. Overemphasis of culture and neglect of civilization is worrying. Civilization is also a process related to time, with past, present, and future trajectories, its space defined by the times. Therefore, talking about culture without discussing civilization is at the very least naive and dangerous. A nation, a country that does not pay attention to its future—the progress of civilization—will be eliminated by history. Therefore, members of a wise nation is always creating cultural traditions, not inheriting them.

Design Is the Third Kind of Wisdom and Ability

In the beginning, design served the purpose of solving problems related to small production methods, not adapting to large production methods. Design has been proven to be the third branch of wisdom and capability contributing to the future of humankind, in addition to science and art. In today's world of economic globalization, technological potential expansion, regionalization of demand, and personalization of consumption, the original tasks of design have changed drastically, and the transformation of human future survival methods begun.

China faces unique problems in terms of modes of living, transportation, eating, and aging. Therefore, I always advocate that we apply the design rationale method to research and innovative practice. This theoretical exploration and practice not only promote the process of sustainable development in China, but also have important referential significance for the development of all emerging countries. Design rationale holds that at different times, in different environments, and under different conditions (external factors), the target system limits the satisfaction of specific people's (groups) needs. The process of studying various limiting factors outside of objects entails both positioning the target system and becomes the basis for evaluating, selecting, and integrating principles, materials, technologies, structures, shapes, and products (internal factors) to create new products or service systems. Understanding external factors serves to correct or revolutionize existing products or humankind's services target system, which is the essence of design. The goal is to do things and solve problems, not to obsess over material product design. The standard for evaluating the innovativeness of a design rationale comes from ancient Chinese philosophy and entails assessing what is appropriate, moderate, and suitable, rather than the infinite pursuit of desire and optimal solutions.

Since the end of the 20th century, I have advocated that service design is the best application field of design rationale and the inevitable direction of contemporary social design. Service design connotes that the fundamental purpose of design is to create a healthy, reasonable, shared, and fair survival method for human society. The history of human civilization's development is a process of continuously changing economy, technology, business, wealth and ethics, morals, and values. The fundamental purpose of service design is not to satisfy the human desire for material and resources, but to serve humans, to solve potential problems pertaining to survival and development. This constitutes a transformation of human civilization from people-oriented to ecology-oriented values. A sharing-type social service design opens the door of hope for the sustainable development of humankind.

China's Solution: The Mission of Design

China is currently facing both the crisis and opportunity of the 21st century. How will the manufacturing industry of the big data era and artificial intelligence develop? What insights will the future international strategic layout and social forms give us?

Our country is facing the dual pressure of the revival of high-end manufacturing industries in developed countries and the diversion of low-end industries in developing countries. The global industrial situation is undergoing profound adjustments and changes. A new international industrial division of labor is coming into existence. Looking at today's global development trends, industrial design is an important aspect of the manufacturing sectors in many developed countries. Its enormous value profoundly affects the development of today's economies, cultures, and societies. Compared with what is occurring in developed countries, the role industrial design plays and its value still have a lot of room for growth.

China's industrial landscape remains in the processing stage of manufacturing, with a significant lack of control over its industrial processes. While there is a robust industry, true industrialization has not been fully achieved. Many so-called designs are mere imitations and adaptations of foreign principles and technologies, largely limited to superficial enhancements of appearance rather than substantial innovations. Current design trends, styles, theories, and methods often reflect a reluctant submission to "capital, business, or technology"—emphasizing dazzling, stylish, and fashionable product designs—or represent ineffective resistance neglecting environmental and human considerations.

Reflecting on the state of Chinese industries reliant on imports, it is evident they remain in a phase of gradual improvement. In contrast, in sectors where foreign dependence has presented significant challenges, China has established its own advanced industrial systems, achieving an international standard. This dichotomy highlights two critical issues: first, reliance on imports and foreign support has limited focus to production and marketing; second, under pressure, the ingenuity of the Chinese people is stimulated, demonstrating our capability for independent innovation.

The Chinese Solution: Social Redesign

The true measure of China's strength lies not only in having our products on global supermarket shelves or major online marketplaces like Amazon and Alibaba, but also in ensuring that these products, protected by our own intellectual property rights, are present in German and American laboratories. Future manufacturing should produce "machines" capable of thinking, speaking, and communicating. We need to reflect on past lessons: we once treated gunpowder as mere entertainment in the form of fireworks, while others turned it into a deadly weapon. Now and in the future, focusing on the five major development concepts—innovation, coordination, green development, openness, and sharing—China must implement an innovation-driven development strategy, carry out supply-side structural reforms, and take on the significant responsibility of achieving the Made in China 2025 goal.

Chinese design should embrace traditional values and the ethics and philosophy of sustainable development. These should become the bedrock of our design industry, emphasizing social dimensions over market demands and consumption. The real need is not just what people want but what they genuinely need, thus defining demand, leading demand, and creating demand. Fashion should not be mistaken for design; fashion is ephemeral. When designing furniture, for instance, the focus should be on creating homes that are people-oriented, rather than merely changing materials or structures. A house is a physical entity, but a home involves people and is a place to live—a sanctuary for the soul.

The Chinese Solution: A Shared Human Destiny

Design should transcend business interests and become a commitment to the sustainable survival and reproduction of humankind. The industrial revolution has initiated a new era, with industrial design as the driving force for innovations in large-scale production. Yet, the current utilitarian industrial economy, driven by sales, profit, and resource domination, has diverted industrial design from its fundamental objective: creating a fair survival for humankind. Today, human desires often eclipse reason and conscience, exacerbating problems like population growth, environmental pollution, resource depletion, and wealth inequality. However, there is a growing awareness that humanity cannot endlessly deplete resources and space needed by future generations.

Chinese design should promote use over possession and focus on creating user experience value rather than merely exchange value. It should go beyond product design to embrace sharing service design. Industrial design is not merely a skill but an innovative model involving cross-border, integrated, leading innovations.

Sustainable design aligns with the development goals of the Chinese Solution, involving the reorganization of knowledge structures, industrial chains, resource integration, and the development of healthy, reasonable, and sustainable human societies.

Advocating for use rather than possession embodies the essence of sustainable design. This approach can inspire humans to pursue simplicity, harmony, and beauty, unleashing the inner potential of human evolution without resorting to luxury and excess, which ultimately lead to corruption and humanity's destruction. We must create a way of life that has never existed before and embark on a path of development unique to China—China's Solution: A Shared Future for Humanity.

Luxury is not a tradition in Chinese culture and will not win global admiration.

Our dream is the rejuvenation of the Chinese nation, as articulated by President Xi articulated it, not the dream of getting rich, nor prosperity enjoyed by a few financial groups that control most of the world's resources and wealth.

This book comes highly recommended. I dedicate it to those aspiring to change the world: designers, engineers, educators, entrepreneurs, and dreamers.

Humanities at Tsinghua University
Beijing, China
November 2021

Liu Guanzhong

Foreword II: The Current Mission of Design

In response to the worsening climate crisis, the United Nations has formulated 17 global sustainable development goals to be achieved by 2030. China has also proposed the strategic task of building an ecological civilization and attaining its carbon peak by 2030 and carbon neutrality by 2060. This sets a clear path for the sustainable development of our national economy and society. Designers, as providers of innovative solutions and constructors of social meaning, need to actively assume their historical missions, make positive contributions, and engage in systematic sustainable theoretical research and design strategy guidance.

As early as 1971, Victor J. Papanek critiqued the excessive commercialization of design in his seminal work, *Design for the Real World.* He posed a crucial question: Why do so many carefully designed products end up damaging the environment? This query underscores the imperative of sustainable design—a process undertaken by the design discipline to deeply reflect on contemporary environmental, social, and economic challenges. Sustainable design has evolved into a pivotal direction and a broad consensus in the realms of future design practice and education.

Over nearly half a century, sustainable design has been recognized and disseminated widely across developed countries. Numerous design schools globally have integrated sustainable design as an independent teaching module, developing relatively systematic pedagogical methods and consistently seeking innovative solutions in collaboration with related disciplines. Amid the transformative opportunities and challenges posed by the integration of new liberal arts and new engineering within China's industrial design education, sustainable design—alongside green and ecological design—is becoming central to design teaching and research.

Driven by both international frontier trends and national development strategies, many domestic design schools in China have incorporated courses related to sustainable design. However, there remains a significant gap in the availability of monographs, textbooks, and reference books dedicated to the theme of sustainable design. The publication of this book you are about to read addresses this gap and is of paramount importance for the teaching and research of design majors.

The content of this book synthesizes advanced theoretical research results from around the world and incorporates a significant amount of original content, reflecting the latest research findings and unique perspectives on sustainable design in China. Sustainable design is the theme of global research and practice that necessitates extensive international exchange and cooperation. The three authors of this book have actively engaged in related international cooperation projects for many years, demonstrating a broad global vision. They have conducted extensive and in-depth collaborative scientific research, teaching, and design practice with international counterparts in developing countries across Asia, Africa, and South America, achieving notable results.

As participants in several international cooperation projects, the authors have had the opportunity to exchange ideas with experts worldwide and contribute to meaningful discussions. Notably, all three authors have been deeply involved in the International Learning Network on Sustainability (LeNS) project initiated by the European Commission.

They have engaged in extensive discussions of numerous practical explorations, experiences, and accumulated knowledge in the field of sustainable design with experts from around the world. This book integrates international perspectives with local practices, offering readers a both comprehensive and systematic knowledge framework and learning manual on sustainable design. It holds significant reference value and provides essential guidance for scholars and practitioners in the field.

Academic Committee of the School of Design Art
Hunan University
Changsha, China

Industrial Design Teaching Guidance Sub-Committee
of Higher Education Institutions, Ministry of Education
Beijing, China
January 2022

He Renke

Foreword III: From Capital-Centered to People-Centered

Former Dean of the School of Agriculture and Rural Development
State Council Expert with Special Allowance

At the beginning of 2021, President Xi Jinping proposed the Three News: grasping a new development stage, implementing a new development concept, and building a new development pattern. This positioning is crucial for the national economy and people's livelihood and serves as a macro background for discussing sustainable design.

Over the past forty years, China has navigated comprehensive industrialization by integrating into the global trade system, albeit at a significant cost. This cost has been multifaceted. We have exploited domestic resources extensively and, when those were insufficient, imported resources on a large scale to support massive production. The resulting goods were then exported at low prices to secure foreign exchange, which was reinvested to further expand production. This cycle of industrial capital expansion has led to severe environmental degradation: resource depletion and the visible destruction of air, water, and soil quality. Concurrently, large-scale production has necessitated cheap labor. To manage rising costs in education and healthcare due to marketization, countless individuals have left their hometowns to seek livelihoods on assembly lines, sacrificing their youth and health.

This scenario represents the traditional capital-driven endogenous growth model during the earlier phase of industrial expansion, reminiscent of the Three News concept. At the turn of the century, as our nation embraced globalization, the primary challenge was capital scarcity. Today, however, the global landscape, our country included, faces a capital surplus. In response, around 2015, we embarked on industrial supply-side reforms to address overproduction. Initiatives such as "three reductions and one drop," industrial shutdowns and conversions, and upgrading facilities signify a shift from extensive development to enhancing industrial production's value and reducing environmental degradation.

Simultaneously, our poverty alleviation strategy aimed to redirect excess production capacity to impoverished areas. The state issued bonds to fund infrastructure development in those regions, enhancing the efficiency of local resource-based production factors and improving living standards. Although those investments may not yield immediate market returns, they align with the long-term benefits of the Two Mountains theory (lucid waters and lush mountains are invaluable assets) and people-centered development. This holistic approach integrates environmental sustainability, poverty reduction, and improved livelihoods, reflecting the new paradigms for development.

Within this macro context, the discipline of design must evolve its focus. Historically centered on urban industrial capital, it should now pivot toward harnessing the value-added benefits of ecological resources through sustainable design. Sustainable design encompasses environmental, social, and economic sustainability. While design may not match the influence of finance or public management, it offers unique advantages. Can designed products and services prioritize environmental and social considerations? Beyond catering to the affluent, can design promote better consumption practices and lifestyles? Can design processes integrate livelihoods and social benefits, contributing to underdeveloped areas and rural revitalization?

Many designers and researchers are already exploring these questions. *Sustainable Design* represents a valuable starting point on this journey. The 2018 Amendment to the Constitution of the People's Republic of China by the National People's Congress underscores the importance of "promoting the coordinated development of material civilization, political civilization, spiritual civilization, social civilization, and ecological civilization." These pillars are essential to the rejuvenation of the Chinese nation, setting the development trajectory for design toward a prosperous, democratic, civilized, harmonious, and beautiful socialist modernized country.

Therefore, I encourage everyone to strive together.

Renmin University of China
Beijing, China
January 2022

Wen Tiejun

Preface

The thinking and writing that have contributed to this book have been taken place over the course of many years. During this period, the issue of design sustainability—in design education, research, and practice—has become increasingly urgent and important. This is now starkly apparent, especially since 2020, when the COVID-19 pandemic highlighted our global interconnectedness and crises. This world is full of uncertainties. The relationship between humans and nature remains an eternal topic. How will design practitioners respond to society's expectations and demands for sustainability? How can individuals update their knowledge and strengthen their abilities to promote a sustainable society? The answers to these questions are central to this book.

The Story of Easter Island

Easter Island, situated in the remote South Pacific more than 3,000 kilometers from the South American continent, is the most remote and isolated inhabited island on Earth. Today, Easter Island is primarily a tourist destination known for its 887 mysterious giant stone statues, known locally as Moai. These statues vary in size and feature peculiar faces with serious expressions facing the sea. Most of the Moai were found destroyed and toppled, with many unfinished statues scattered across the island, suggesting a history of civilization and conflict, prosperity, and decline.

Easter Island is a volcanic island covering about 166 square kilometers. It once had fertile land and dense forests, with an excellent climate, making it a potential paradise. Archaeological discoveries have confirmed the existence of a once-thriving civilization. Between the 3rd and 5th centuries AD, the Rapa Nui people began to inhabit the island, gradually forming numerous tribes. As life prospered, tribes competed to build stone statues and sacrificial platforms for religious ceremonies. At its peak, the island's population reached 15,000, far exceeding its ecological capacity, which scientific calculations estimate is a maximum of 2,000 people.

By around 800 AD, a population explosion led to almost complete deforestation, the near extinction of animals and birds, and severe damage to the ecology of the shallow sea surrounding the island. The Rapa Nui people transitioned from fishermen to farmers, raising chickens and growing crops like sweet potatoes, taro, and sugarcane. However, their declining yields could not sustain the population. In desperation, they resorted to cannibalism,[1] and their once-complex social structure collapsed. By around 1700, after experiencing starvation, violence, and chaos, only about 2,000 people remained on the island. In the 19th century, the surviving Rapa Nui faced large-scale abductions by Western slave traders and devastating epidemics. By 1877, only 110 people remained on Easter Island.[2]

The rise and fall of Easter Island illustrate the consequences when societal and economic development surpass a region's environmental and resource-carrying capacity, leading to civilizational decline. The islanders, unaware of their isolation and dependence on limited resources, consumed indiscriminately, resulting in their civilization's brief prosperity and subsequent collapse.[3] During the course of human history, there have been numerous, similar examples.

Indeed, Earth can be seen as a large version of Easter Island. For millennia, humans have developed resources without restraint so as to build highly "civilized" and prosperous societies. However, the planet's resources are finite, and once exhausted, disaster is inevitable and there will be no escape. Fundamentally, most historical conflicts and wars are energy struggles, exacerbated by human greed and strife. We inhabit a lonely spaceship[4] in the vast universe, without external aid or the opportunity for resupply. The question arises: will the human world repeat the fate of Easter Island? This is a question upon which each of us must critically reflect.

---

[1] Archaeological studies of Easter Island revealed significant amounts of human bones in garbage produced during the late period of occupation. Although Polynesians in other places are reputed to engage in cannibalism, this is mostly for religious or superstitious reasons and only occurs on special occasions. But the cannibalism of the Rapa Nui people had a very practical purpose: to supplement protein consumption. Reference: Reese. Easter Island: The Disappeared World and Mysterious Civilization [M]. Xu Qunhang, Chai Junyang, trans. Shanghai: Shanghai Science and Technology Literature Publishing House, 2017: 32-35.

[2] Reese. Easter Island: The Disappeared World and Mysterious Civilization [M]. Xu Qunhang, Chai Junyang, trans. Shanghai: Shanghai Science and Technology Literature Publishing House, 2017: 68.

[3] Ma Zhong. Introduction to Environmental and Natural Resource Economics [M]. Beijing: Higher Education Press, 2019:3.

[4] American economist Henry George published *Progress and Poverty* in 1879, which viewed the earth as a spaceship with limited resources; George Orwell provided an illustration of Henry George's theory in *The Road to Wigan Pier* (1937). In 1966, renowned economist Kenneth Boulding published *The Economics of the Coming Spaceship Earth*, using George's theory to explore how humanity might develop under the condition of a limited total inventory. By the late 1960s, a spaceship theory had formed, with far-reaching influence.

The Revelation of Several Numbers

Our planet, often called Spaceship Earth, is remarkably old at 4.6 billion years. The Gaia Hypothesis[5] portrays Earth as a dynamic entity akin to a living organism, embodying the essential characteristics of life. Despite initial skepticism among scientists, the Gaia Hypothesis transcends philosophical conjecture, as the intricate and continuous interplay between life forms and the natural environment evidence. This symbiotic relationship maintains a delicate equilibrium akin to a finely tuned life system. It is crucial to recognize that humans neither reign as masters nor stewards of the Earth but rather coexist as one of Mother Earth's multiple progeny.

Human history spans a mere 200,000 years—a fleeting moment in Earth's vast timeline. As biological entities inhabiting Mother Earth, humans possess unparalleled intelligence and creativity. Until a certain point in the course of evolution, a harmonious coexistence prevailed between humans and nature. However, over time, humans transitioned from cohabitating with other creatures to envisioning total domination over the planet.

The Industrial Revolution, spanning over two centuries, has bestowed upon us material prosperity inconceivable to our ancestors. However, it has also initiated disruptions to the planet's pristine ecological balance from cradle to cradle. While advancing civilization, humans have contaminated the air, the soil, and water sources. Rapid advancement propelled by science and technology has surpassed even our own imaginations, leading to a loss of control over our impact on the environment.

China's forty-year journey of reform and opening up has garnered global acclaim for its developmental achievements, yet the toll exacted has been substantial. Transitioning from a model of extensive development to one of ecological civilization and sustainability will pose a formidable challenge and a paramount mission to future generations of Chinese society.

Human activities have left an indelible mark on the planet, even penetrating geological depths. In the latter half of the 20th century, there was an exponential acceleration of activities deemed beyond control, marking an unprecedented era in human history. The past five decades have ushered in the most rapid transformation in the human–nature relationship,[6] propelling us into what experts term the Anthropocene—an epoch where human activities create changes in the Earth's systems.[7]

---

[5] Gaia is the name of the earth goddess in Greek mythology. James Lovelock proposed the Gaia hypothesis, viewing the earth as a living entity and defining Gaia as a complex entity involving the earth's biosphere, atmosphere, oceans, and soil that seeks the best physical and chemical environment for life on earth and maintains relatively constant conditions through active control.

[6] Steffen W, Sanderson A, Tyson P D et al. Global Change and the Earth System: A Planet Under Pressure. Berlin: Springer, 2004: 131.

[7] To emphasize humans' central role in geology and ecology today, Nobel laureate in Chemistry Paul Crutzen proposed the concept of the Anthropocene. Since the industrial revolution in the mid-late 18th century, humans have become a significant force influencing environmental evolution,

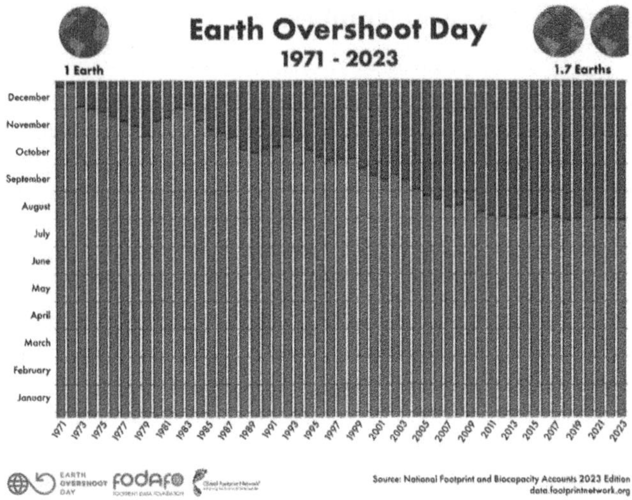

Fig. 1 Earth Overshoot Day data (Earth Overshoot Day denotes the date when we (all of humanity) have used more from nature than our planet can renew in the entire year. The data in the figure shows that humans have been overusing Earth's ecological resources since 1970. The overshoot day in 2021 was July 2. See: https://www.footprintnetwork.org/our-work/ecological-footprint/)

According to data from the Global Footprint Network (2021), human consumption has already exceeded the Earth's ecological carrying capacity at the equivalence of 1.7 Earths.[8] Should current trends persist, projections indicate by 2050, we will require the resources and capacity of 2.9 Earths to sustain our development and mitigate our waste.[9] However, we inhabit only one Earth, which renders this excess consumption a borrowing from future generations—a debt upon which we are destined to default. Research from the World Business Council for Sustainable Development (WBCSD) underscores the urgent need for industrialized societies to curtail resource consumption by 90% within the next 50 years for us to have hope of achieving sustainability.[10] Do you have confidence in our abilities to do this? (Fig. 1)

---

especially in the past century. Meanwhile, the speed of urbanization has increased tenfold and several generations have depleted fossil fuels that took millions of years to form. In the future, even for 50,000 years, humans will still be a major geological driving force. Therefore, it is necessary to study the earth system from the new perspective of the Anthropocene, paying attention to the huge, unignorable impact that humans have had and will continue to have on the earth system. Reference: Crutzen P J, Stoermer E F. The "Anthropocene"[J]. IGBP Newsletter, 2000(41): 17-18.

[8] The data comes from the official website of the Global Footprint Network: https://www.footprintnetwork.org/our-work/ecological-footprint/

[9] World Wildlife Fund. Living Planet Report 2012 Special Summary [EB/OL]. [2021-09-26]. https://www.wwfchina.org/ content/press/publication/2012lpr-riocn.pdf

[10] WBCSD. Eco-Efficient Leadership for Improved Economic and Environmental Performance[R/OL]. 1996[2021-09-26]. http://wbcsdservers.org/wbcsdpublications/cd_files/datas/wbcsd/business_role/pdf/eeleadershipforimprovedeconomic&enviperformance.pdf

It can be imagined that if human society continues to develop at its current pace of growth, in the next 50, 100, 200 years, we may achieve even more "brilliant" material achievements. But, at the same time, our environmental impact will increase, as will our grave social and economic problems. The question will no longer be one of sustainable development, but likely one of human survival. The tragedy of Easter Island is not far in the future.

Design and Sustainability

Herbert Alexander Simon posits that any endeavor aimed at projecting knowledge, experience, and intuition into the future to alter the status quo possesses a design essence. Simon thus construes design as a purposeful, forward-thinking creative pursuit that humans undertake. Simply stated, design bears the dual responsibility of addressing immediate challenges or imbuing significance in the present, while also envisioning and shaping a trajectory for the future.

Since antiquity, human creative endeavors, including contemporary industrial product design, have revolved around identifying and resolving problems, as well as engendering value or significance. The essence of design itself remains constant, with only the nature of the problems it solves and its meanings in flux. Historically, factors such as form, function, usability, commercial appeal, brand identity, and manufacturing efficiency held significance, while designers' tasks included enhancing product aesthetics, suitability, ease of production and transportation, and fostering distinctive brand images. Now, the landscape of design is more intricate than ever, with pressing issues such as environmental degradation, resource scarcity, food security, public health, social equity, cultural diversity, urban–rural dynamics, and aging populations demanding attention. These multifaceted challenges underscore the imperative for designers to confront complex sustainable development issues head-on.[11]

Moreover, beyond addressing current realities, envisioning the future constitutes a pivotal responsibility for designers. Design's goals include articulating aspirations and dreams, portraying a vision of an improved future existence, and illustrating numerous paths that could lead us a sustainable society. Indeed, design inherently possesses a future-oriented dimension. In the face of an uncertain future, designers anticipate technological advances, societal expectations, potential conflicts, and challenges, while advocating for specific positions to shape a collectively envisioned future—whether through scenarios, systems, structures, services, or products. Design itself is never an end goal but rather a conduit for achieving valuable objectives. Thus, within the context of contemporary social transformation, whether tackling present or future challenges of significance, sustainability emerges as the overarching aim of design.

[11] Liu Xin. The Boundary of Design[J]. Decoration, 2014 (Supplement): 134-136.

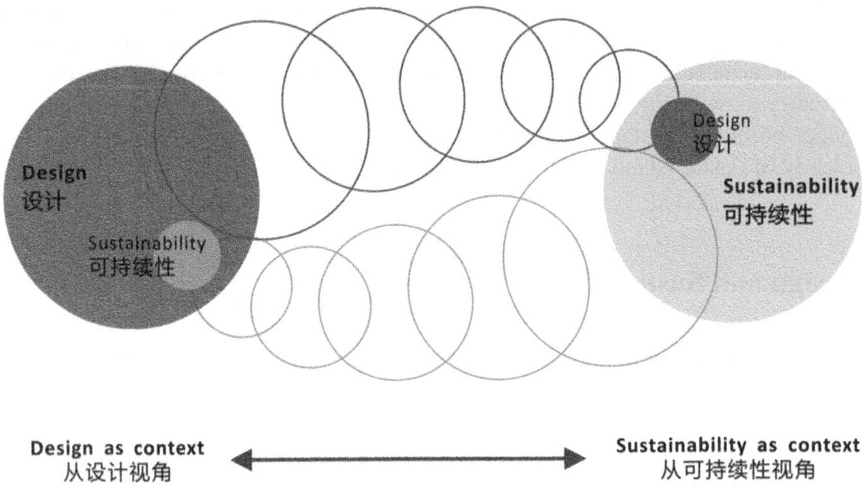

Fig. 2 Relationship between Design and Sustainability (Dewberry E, Fletcher K. Demi: Linking Design with Sustainability[C/OL]. The 7th European Roundtable On Cleaner Production, IIIEE, May 2-4, 2001, Lund, Sweden. 2001[2022-03-10]. http://libeprints.open.ac.uk/27047/1/ercp_lund_2001.pdf)

Sustainability long occupied a moral high ground or was relegated to a seemingly peripheral research focus within the field of design. However, at present, sustainable development has surged to the forefront as humanity's paramount concern, with design intricately interwoven into its fabric, serving as one among numerous instruments for realizing sustainability goals. This evolution in the role of design is palpable in the successive redefinitions of industrial design (Fig. 2).[12]

As we discuss them today, design and sustainability are closely related. To some extent, good design is sustainable design. As design practitioners, we need to step out of the limited perspective of traditional design professions and understand the role and value of design from a broad perspective of environmental, social, cultural, and economic sustainable development. This is the foundation of sustainable design.

[12] Since the 1950s, the definition of industrial design has undergone five revisions. In 2015, the International Council of Societies of Industrial Design was officially renamed the World Design Organization and announced the latest definition of industrial design: [Industrial] design aims to guide innovation, stimulate business success and provide a better quality of life. It is a professional interdisciplinary activity that applies strategic problem-solving processes to the design of products, systems, services, and experiences. It closely links innovation, technology, business, research, and consumers in creative activities, visualizes the problems to be solved and the proposed solutions, deconstructs the problems, and uses them as opportunities to establish better products, systems, services, experiences, or business networks, providing new value and competitive advantages. (Industrial) design responds to social, economic, environmental, and ethical issues through its outputs, aiming to create a better world. Reference: https://wdo.org/glossary/industrial-design/

Active Activists

Despite the formidable challenges and uncertainties that loom over the future, a glimmer of hope remains. Because "in addition to self-harm, human nature also has the ability to love and perceive, and may use wisdom to find ways to reduce self-harm….Moreover, we can only be full of confidence, foreseeing future disaster scenarios will help us give birth to collective wisdom, just as understanding death can help a person live more wisely (and cherish more)."[13] Indeed, envisioning potential disaster scenarios can galvanize collective wisdom, akin to how contemplation of mortality fosters wiser living and greater appreciation for life. Undoubtedly, the realization of a more sustainable world necessitates a profound transformation of current lifestyles, production methods, and economic models. However, time is of the essence, and the window for effecting these changes is rapidly closing.

Unfortunately, consensus remains elusive regarding attitudes towards the future and the actions we need to take. In what follows, in each chapter, we will explore divergent viewpoints, including contrasting paradigms of strong sustainability versus weak sustainability, as well as perspectives rooted in anthropocentrism, technocentrism, and ecocentrism. Broadly speaking, individuals tend to adopt either pessimistic or optimistic stances towards the future, though many may fall somewhere in between or exhibit indifference altogether.

Pessimists envision an irreversible trajectory marked by climate anomalies, pollution, resource depletion, and ecological degradation, akin to the prophesied demise of Easter Island or the dystopian portrayal in Christopher Nolan's film *Interstellar*. Conversely, optimists maintain that technological progress holds the key to resolving the challenges we face. They tout the potential of big data, artificial intelligence, and biotechnology for addressing a myriad of societal issues, from transportation to healthcare to environmental stewardship.

However, while technology undoubtedly wields transformative power, it is not a panacea. Sustainable social transformation necessitates nuanced changes across social, cultural, and economic systems, transcending technological solutions. Moreover, technology is neither neutral nor isolated; it operates within the constraints of business interests, ideologies, political systems, and cultural norms. The full realization of technological capabilities will require time, experimentation, iteration, and societal adaptation—none of which are readily achieved overnight.

Furthermore, skeptics may dismiss sustainability as a human construct divorced from the inherent rhythms of the planet. They argue that Earth's natural processes will persist regardless of human intervention, rendering sustainability a mere projection of human desires rather than a fundamental aspect of the planet's existence. Acknowledging the humility inherent in this perspective underscores the inextricable link between human civilization and the natural world.

Ultimately, the future of the planet remains uncertain. Yet, we persist in yearning for a better tomorrow—a world where human achievements harmonize seamlessly

[13] Carlo V, Ezio M. Design For Environmental Sustainability[M]. London: Springer, 2008: 3.

with a flourishing, unspoiled natural environment. Realizing this vision will demand collective effort from all corners of society. As stewards of design, practitioners bear a profound responsibility to contribute towards this shared goal. Regardless of our stance regarding the future, we designers serve as proactive agents of change, continuously innovating and advocating for solutions that pave the way towards a sustainable society.

Tsinghua Garden
Beijing, China
2021

Liu Xin

Acknowledgment

This collaborative effort involved multiple authors, including doctoral students from Tsinghua University and Dr. Maurizio Vrenna from Politecnico di Torino, Italy. The contributors, all members of the Learning Network on Sustainability (LeNS), drew upon their extensive research, teaching, and practical experiences to enrich the book's content.

We are grateful to Professor Liu Guanzhong, Professor He Renke, and Professor Wen Tiejun for contributing their prefaces to the Chinese edition. Their feedback and support is a treasure trove of insights for us.

Upon the publication of the English edition, we would like to extend our gratitude to Professor Qiu Song and Associate Professor Zhong Zhou for their assistance in liaising with Springer. We also thank Mr. Zhang He from Springer Nature China for his support throughout the publishing process. Special thanks go to Professor Ezio Manzini, Emeritus Professor at Politecnico di Milano and founder of the DESIS network, and Professor Carlo Vezzoli, initiator of the LeNS network, for contributing the foreword to this book. Additionally, we are deeply appreciative of Dr. Huang Yujun for his invaluable help in proofreading the manuscript. Our heartfelt thanks go to all of them.

Guided by a commitment to academic rigor and accessibility, we seek to inspire critical thinking and action within our readership. Recognizing the inherent complexities and uncertainties surrounding sustainability discussions, the authors eschew the notion of providing definitive standard answers, opting instead to foster contemplation and engagement. Imperfections inherent in the text are acknowledged, with gratitude extended to readers for their understanding and patience as part of this ongoing learning process.

Competing Interests The authors have no competing interests to declare that are relevant to the content of this manuscript.

About This Book

The core of this book revolves around the theme of design. Therefore, our central focus lies in exploring how we can harness design theory, methods, and strategies to forge a sustainable future. Through meticulous data collection, analysis, reflection, and refinement, we have meticulously crafted 12 chapters broadly categorizable into two sections: theory and strategy.

The first two chapters comprise the theoretical segment. They delve into the historical background, contextual development, and key viewpoints—along with associated debates—that pertain to sustainable development and design. These chapters aim to provide readers with a macro-level understanding of the field's evolution.

Beginning with Chap. *3*, the discussion explores various design strategies in detail. Chapters *3* to *8* predominantly explore the environmental dimensions of design strategies, encompassing themes such as lifecycle design, circular regeneration design, reuse and upgrade reconstruction design, durability and emotional design, reduction design, and biomimicry design. While these strategies inherently entail considerations of complex systems that include social, cultural, and psychological elements, their primary focus is environmental sustainability.

The topic of Chap. *9* is sustainable product-service system design, a distinctive aspect of sustainable design strategies that transcends traditional product-centric approaches. This branch of design integrates products and services into a holistic solution by including considerations of economic models and consumption patterns. Social innovation design, representing a frontier realm of sustainable design research characterized by multifaceted human-centric challenges, takes center stage in Chap. *10*. Chapter *11* explores distributed systems and sustainable design, offering insights into potential future economic models that offer novel perspectives on sustainable transformation. Integrated system design, discussed in Chap. *12*, entails the comprehensive application of the above strategies across varying scales and levels. The book concludes with a comprehensive review and summary of its key insights and findings.

Contents

Chapter 1
Sustainable Development

Abstract Sustainable development emerged in response to increasingly intense conflicts between human society, economic development, resources, and the environment. While discussions about the relationship between humans and nature have a long history, this topic became especially prominent after the Industrial Revolution, due to the intensification of environmental problems. The United Nations (UN) officially introduced the concept of sustainable development in the 1980s. It encompasses three main characteristics: the efficient use of natural resources to maintain and promote a healthy ecological environment, economic development as a requirement, and the aim of comprehensive social progress.

Keywords Nature history · Sustainable development · Ecological footprint · Anthropocentrism · Technocentrism · Ecocentrism

Although sustainable development as an idea originated in the West, in recent decades, the conflicts it aims to mitigate have become increasingly prominent in China, with rapid economic development exacerbating them. However, the concept of sustainability is also deeply rooted in Chinese culture. Traditional Chinese philosophy emphasizes the unity of humans and nature, adherence to natural laws, adapting to local conditions, making optimal use of resources and exercising restraint. Integrating Western sustainable development concepts into Eastern traditional wisdom to form a shared knowledge system and spiritual wealth for contemporary society is a challenge design scholars are currently addressing. This integration aims to support sustainable social transformation and the construction of a global community with a shared future.

Views on sustainable development vary widely in both academic and public discourse. Some equate it with radical environmental protection, with European scholars even advocating for degrowth to minimize human impact on the natural environment. Others emphasize the social aspect, focusing on the basic survival and welfare of low-income groups, and critique environmental protection measures that lack differentiated responsibilities and economic development that disregards the principle of fairness.

X. Liu et al., *Design for Sustainability*, Sustainability and Interdisciplinarity,
https://doi.org/10.1007/978-981-96-8657-5_1

China's situation is particularly complex. The socioeconomic development of the eastern coastal areas has reached the level of moderately developed countries, while consumerism strongly attracts the younger generation. Simultaneously, although China has recently achieved significant poverty alleviation, a large portion of the population remains in the middle- and low-income brackets, with a substantial urban–rural divide. Moreover, issues such as climate change, ecological crises, environmental pollution, resource scarcity, and various social contradictions are pressing. These challenges compel China to seek a sustainable development path tailored to its unique national conditions.

To anticipate the future, we must review the past. This chapter traces the historical origins and various development models of human society, aiming to help readers understand the evolution of sustainable development concepts and their critical importance and urgency in today's world.

1.1 Interaction Between Humans and Nature

1.1.1 From Prehistory to Agricultural Society

Earth is humans' only home. Before the appearance of homo sapiens, countless lifeforms inhabited our planet. With changes in Earth's climate, environment, and geology, new species emerged and went extinct. From trilobites to dinosaurs, various creatures have appeared and disappeared on Earth, while the planet continues its orbit around the sun, seemingly an eternal existence (Fig. 1.1).

On our planet, the largest ecosystem is the biosphere, where all life resides and reproduces. Various animals, plants, and microorganisms obtain life-sustaining nutrients from the environment. These nutrients, after excretion, decay, and transformation, return to the environment, stored once more in non-biological mineral form. In this system, various substances circulate, neither increasing nor decreasing in quantities. In this vast biosphere, countless ecosystems contain actors playing three basic roles: producers (such as plants), consumers (such as animals), and decomposers (such as bacteria, fungi, and other microorganisms). Plants absorb nutrients from the atmosphere and soil, converting solar energy into plant tissue through primary production, which forms the basis for the survival of all consumers, including humans.

Humans are not at the top of the food chain due to the presence of many natural enemies and we are not natural rulers. However, through socialization and cooperation, humans have developed unique communities and, based on this, developed culture through collective learning.[1] Through culture, rather than pure natural evolution, we have adapted to the environment faster and better than other species,

[1] Heinrich. *The Secret of Our Success: How Culture Is Driving Human Evolution, Domesticating Our Species, and Making Us Smarter* [M]. Beijing: CITIC Publishing Group, 2018:

Fig. 1.1 Earth photo taken by China's Fengyun-4

gradually assuming a dominant position in the biosphere.[2] Over time, humans have freed themselves from the constraints of nature and embarked on a journey to change and rule the world.

Approximately 200,000 years ago, our ancestors, homo sapiens, left Africa and began to journey to every corner of the planet. Their ability to use tools and fire, and to live in groups, allowed them to have a significant impact on their environments. It is estimated that 34 of the 47 genera of large mammals originally in North America disappeared within two thousand years after humans arrived. Paleontological research indicates that the mass extinctions in Australia and the Americas coincide with the arrival of humans. Our ancestors, tens of thousands of years ago, had already profoundly changed the Earth's ecological environment.[3]

With the advent of agricultural society, humans improved energy conversion efficiency by domesticating animals and plants. This led to increased complexity in human society and culture, the emergence of high-level civilizations, and slow population growth. However, agriculture also had its pitfalls. According to Green World History, in fertile Mesopotamia, agricultural productivity peaked but then began to decline. Reports of "land turning white" around 2000 BC show the consequences of

[2] Christian. *Origin Story: A Big History of Everything*[M]. Sun Yue, trans. Beijing: CITIC Publishing Group, 2019: 204.

[3] Harari. Sapiens: A Brief History of HumankindA[M]. Lin Junhong, trans. Beijing: CITIC Publishing Group, 2017: 66.

salinization due to over-cultivation. By 1800 BC, grain production had decreased to only a third of its previous output, leading to the collapse of Sumerian agriculture, which was then conquered by Babylon and gradually disappeared.[4]

The development of agriculture, population growth, and environmental degradation occurred in nearly all early civilizations. In the Mediterranean, the natural vegetation was once a mixed forest of evergreen and deciduous trees. Today, however, it is known for olives, grapes, and herbs due to deforestation, overgrazing, and agricultural cultivation, leading to soil fertility decline and environmental degradation. Plato recorded this change in his *Critias*.[5] In China, the silt in the Yellow River results from land erosion after the destruction of upstream vegetation. The *Book of Songs* records that in the southern part of the Loess Plateau, there were herds of wild deer, tigers and leopards, and lush forests. However, population growth, large-scale construction, deforestation, and intensive farming led to vegetation degradation.[6] By the early days of the People's Republic of China, forest coverage was only 6.1%.[7]

Despite these environmental effects, human destruction of the natural environment was limited at that time. Even in the most developed agricultural regions, the environmental impact humans caused, while severe locally, was negligible overall. Compared to today, it was still an empty world,[8] and nature had sufficient space and time to recover.

1.1.2 From the Age of Discovery to the Industrial Revolution

With the growth of human abilities and the broadening of our horizons, our innate desire to explore the unknown and acquire land and wealth also grew. In 1492, Columbus embarked on his famous journey to the east, marking the beginning of the Age of Discovery and ushering Europe into a phase of large-scale colonial development. When Spanish colonists arrived in the Americas, they cleared wilderness, burned forests, exterminated indigenous populations, and planted large expanses of land with crops economically valuable in Europe, such as sugarcane, tobacco, and corn. This drastically altered the continent's original ecosystem and

[4] Ponting. *A New Green History of the World: The Environment and the Collapse of Great Civilisations* [M]. Wang Yi, trans. Beijing: China University of Political Science and Law Press, 2015: 60.

[5] Same as 4: 64.

[6] Elvin. The Retreat of the Elephants: An Environmental History of China [M]. Mei Xueqin, Mao Lixia, Wang Yushan, trans. Nanjing: Jiangsu People's Publishing House, 2014: 26.

[7] Huang Yiping. "The Loess Plateau's Ecological Environment Has Undergone Tremendous Changes in Seventy Years[N]." China Science Daily, 2019-09-03 (8).

[8] Wei Bole, Vijay K. Man. Flipping the Limit: The Path to the Awakening of Ecological Civilization [M]. Translated by Cheng Yiheng. Shanghai: Tongji University Press, 2019: viii.

initiated what has been termed the "great integration in the history of world flora and fauna."[9]

When Europeans arrived in the Americas and Australia, they hunted extensively, first for sustenance and later to meet the demand for luxury goods among the wealthy. This led to the extinction of numerous animal and plant species.[10] Colonists also introduced various animals and plants, intentionally or unintentionally. For instance, British farmers brought rabbits to Australia, which created a permanent ecological imbalance and led to the extinction or reduction of many native species. As a result, ecosystems in various geographical regions became increasingly similar, resulting in a loss of biodiversity.[11]

Colonizers inflicted long-term, widespread damage on local agricultural systems. Indigenous inhabitants of South America and Southeast Asia had developed over thousands of years traditional agricultural models suitable for their local climates, species, and conditions. However, colonizers prioritized economic gains, focusing on large-scale cultivation of cash crops for low-cost export to Europe. To quickly maximize profits, colonizers abandoned traditional agricultural techniques such as intercropping, crop rotation, and fallowing. This led to soil degradation, loss of fertility, proliferation of pests and diseases, and ultimately ecological collapse. Large-scale farming also required cheap labor, fed by the slave trade from Africa.[12] African slave trade, the Americas and Australia, and European markets became closely linked, forming the first global trade network.[13]

Europe also experienced severe ecological damage. The resource scarcity population growth brought about on the European continent was comparable to, or even more severe than, that in China.[14] Around 1700, the mining industry in Saxony,

---

[9] Ponting. *A New Green History of the World: The Environment and the Collapse of Great Civilisations* [M]. Translated by Wang Yi. Beijing: China University of Political Science and Law Press, 2015:125.

[10] In 1869, a region in Brazil exported 170,000 dead birds for feather plucking; the orchid fever in nineteenth-century Europe led to over 100,000 plants being dug out of the Brazilian rainforest each year for export; in the 1970s, North Americans began turning bison into commercial leather, with up to three million bison slaughtered each year, leading to their rapid extinction by the 1990s; the sandalwood trees in Fiji were all cut down between 1804 and 1809, and those in the Hawaiian Islands were cut down by 1825. See: Pang Ting. Green World History: The Decline of Environment and Great Civilizations [M]. Translated by Wang Yi. Beijing: China University of Political Science and Law Press, 2015:133.

[11] Crosby. *Ecological Imperialism: The Biological Expansion of Europe, 900–1900* [M]. Translated by Zhang Xiguo. Beijing: The Commercial Press, 2017:183.

[12] Ponting. *A New Green History of the World: The Environment and the Collapse of Great Civilisations* [M]. Translated by Wang Yi. Beijing: China University of Political Science and Law Press, 2015:158.

[13] Ma Libo. The Origins of the Modern World: A Global and Environmental Narrative from the 15th to the twenty-first Century [M]. Translated by Xia Jiguo. Beijing: The Commercial Press, 2017:86. Robert B. Marks, The Origins of the Modern World: A Global and Environmental Narrative from the Fifteenth to the Twenty-First Century.

[14] Same as 13:187.

Germany, led to unsustainable deforestation for ore smelting within just over a decade. The resulting timber shortage caused many mines to go bankrupt or to close. Carlowitz, a mining inspector at the time, was the first to recognize the consequences of over-exploitation and, in his work *Sylvicultura Oeconomica*, criticized the short-sighted, profit-driven strategies of timber resource use. He first proposed the concept of sustainability. However, after the Industrial Revolution began in the 1760s, Britain's coal production surged, Western European countries entered the industrial age, and resource shortages were no longer a concern. Carlowitz's warnings were largely ignored until 100 years later when the value of sustainable thinking was rediscovered.

The Industrial Revolution ushered Europe into the machine age. The use of steam engines led to rapid increases in coal mining, a drop in prices, and the swift development of other industries, such as textiles, smelting, and transportation.[15] Fossil energy began to replace bioenergy, spurring rapid industrial development and widespread environmental pollution. Between 1780 and 1880, Britain used coal energy to establish the world's most technologically advanced and prosperous economy, simultaneously causing the most severe air pollution in British history. Domestic sewage and industrial wastewater were directly discharged into rivers, leading to serious river pollution and the infamous Great Stink of London in 1858.[16]

In 1872, Robert Angus Smith first coined the term acid rain to describe the precipitation acidification caused by industrial pollution. The direct consequences of environmental pollution and ecological degradation included the increasing prevalence of diseases. Between 1831 and 1866, the UK experienced four large-scale cholera outbreaks.

During the London smog event in 1952, more than 4000 people died in four days, making it one of the most severe environmental pollution events in modern history. On December 5, 1952, a thick yellow cloud descended on London due to a reaction between two by-products of coal combustion, sulfur dioxide and nitrogen dioxide, combined with London's cold fog, making the smog toxic (Fig. 1.2). This event highlighted the severe consequences of industrial pollution and underscored the need for sustainable environmental practices.

From the emergence of modern humans to the present day, whether in the snowy Arctic or the hot and humid tropics, humans have established highly developed, self-sufficient civilizations. These civilizations have prospered on survival and

[15] Kenneth Pomeranz, *The Great Divergence: China, Europe, and the Making of the Modern World Economy* [M]. Translated by Shi Jianyun. Nanjing: Jiangsu People's Publishing House, 2004: 129. Same as 14:187.

[16] During the British Industrial Revolution, the cotton textile industry was the first to mechanize. The wastewater produced in textile factories was discharged straight into rivers, causing pollution. The development of the textile industry drove the development of the chemical dyeing industry. In 1791, Claude Berthollet published *Dyeing Technology* and soon invented the technology to convert chlorine into industrial bleach. The widespread use of chemical dyeing technology increased direct damage to the environment and water resources. The year 1858 was known as the Great Stink of the Thames. Reference: Clapham. Modern British Economic History (Middle) [M]. Yao Zengyi, trans. Beijing: The Commercial Press, 2014.

Fig. 1.2 On December 5, 1952, a thick yellow cloud descended upon London. This was caused by the reaction between two by-products of coal combustion, sulfur dioxide and nitrogen dioxide, combined with London's cold fog, making the smog toxic. (Image source: https://www.history.co.uk)

reproduction, division of labor and cooperation, and the invention and creation of tools, all based on and constrained by nature. However, the rise of global-scale colonization and expansion by Western European countries during the Age of Exploration disrupted this process. In just a few hundred years, colonization altered the ecological balance of the Earth. The colonializing countries quickly accumulated wealth and their societies modernized due to high energy consumption. In contrast, colonized peoples and lands have largely been unable to recover from the consequences, trapped in a cycle of economic and environmental poverty. This situation is not merely the result of anthropocentrism,[17] but rather the inevitable product of capital-centrism. While a small number of people enjoy the benefits of development, all of humanity bears the cost of environmental destruction.

1.1.3 The Great Acceleration

After World War II, issues of ecological destruction and environmental pollution increasingly gained public attention. Significant events include the Los Angeles smog incident of 1954–1955, resulting in over 4,000 deaths; the London toxic fog of 1952, mentioned above; the Kumamoto Minamata disease outbreak in Japan in 1956; the Santa Barbara oil spill in the United States in 1969; the Bhopal chemical plant explosion in India in 1984, which resulted in 575,000 deaths and injured an additional 200,000 people; and the Chornobyl nuclear disaster in 1986. The advent of radio technology and household televisions brought these events directly into public view, eliciting strong social reactions.

From the mid-20th century onward, various data have shown that the rapid development of human society and the economy has profoundly affected the planet's

[17] Anthropocentrism is detailed below.

ecological balance, with increasingly intense consequences. Particularly concerning is global warming, which poses severe threats to the future of humanity. Over the past century, reliance on fossil fuels such as coal and oil has generated significant emissions of carbon dioxide (CO_2) and other greenhouse gases, contributing to global warming. The average global temperature in the 20th century rose by about 0.6 degrees Celsius. The spring thaw in the Northern Hemisphere in spring is now 9 days earlier than it was 150 years ago, and the onset of autumn frost is about 10 days later.[18] Due to global warming, permafrost in the Arctic Circle has decreased by 95% over the past 35 years. These changes will have irreversible impacts on the global environment, including more frequent climate disasters, rising sea levels, epidemics, food crises, and the migration and extinction of species. The great acceleration of environmental changes introduces significant uncertainty to human society, exacerbating conflicts and disputes among nations, countries, and regions.

1.2 From the Environmental Movement to Sustainable Development

1.2.1 The Rise of the Environmental Movement

As noted in the preface, humans possess rationality and the ability to reflect. When faced with impending crises, they can make informed decisions and engage in acts of self-preservation. Indeed, beginning in the 1960s, various environmental movements have emerged, significantly contributing to the revival of sustainable thinking. Key factors driving this revival include significant advancements in scientific research on pollution, particularly studies on radioactive dust, smoke, and chemicals, and their impact on humans and ecosystems. Pioneers such as Rachel Carson[19] and Barry Commoner[20] have translated these specialized findings into accessible books, facilitating widespread dissemination (Fig. 1.3). Additionally, concurrent social movements, including anti-colonialism, the protests against the Vietnam War,

[18] See the World Wildlife Fund website https://arcticwwf.org/places/last-ice-area/

[19] American marine biologist and writer Rachel Carson published *Silent Spring* in 1962. Hers was the first book to reflect on the environmental pollution and ecological destruction excessive use of chemical pesticides and fertilizers cause, and the major disasters it could ultimately bring to humans. It is considered the cornerstone of the world environmental movement.

[20] Barry Commoner published *The Closing Circle* in 1974, which for the first time linked nature, humans, and technology, revealing from an ecological perspective that the root of the environmental crisis lies in the actions and reactions between the artificial technological sphere and the natural ecological sphere, proposing the famous Four Laws of Ecology. We mention this book many times in subsequent chapters. Reference: Krier J E. The Political Economy of Barry Commoner [J]. Environmental Law, 1990 (20): 13.

Fig. 1.3 Rachel Carson and her book, *Silent Spring*. (Image source: https://www.baidu.com)

the feminist movement, and the civil rights movement, have intersected and influenced the propagation of environmental values, garnering attention from diverse sectors of society.

The economic recovery that followed World War II led to the emergence of a substantial middle class in developed Western countries, accompanied by a rise in the general public's level of educational attainment. This context set the stage for the awakening of public consciousness regarding environmental issues. The publication of Rachel Carson's *Silent Spring* in 1962 marked the beginning of the environmental movement in the United States. The book's revelations about the detrimental effects of pesticides on birds, and ultimately on humans, caused significant concern among the American middle class. Carson's work highlighted the disturbing quiet of a spring season devoid of birdsong due to pesticide-induced deaths, underscoring the broad environmental and human health impacts of such chemicals.

Silent Spring elicited a strong public response and caused widespread discussions, leading to legislative and organizational changes. In 1969, the US Congress passed the National Environmental Policy Act (NEPA) and established the Environmental Protection Agency (EPA). This was followed by the enactment of the Clean Air Act and the Clean Water Act. On April 22, 1970, US Senator Gaylord Nelson initiated the first Earth Day event, mobilizing over 20 million people in marches and rallies in New York, Washington, DC, San Francisco, and other cities. This event had a profound impact on the global environmental protection movement.

The environmental movement reached a climax with the founding of Greenpeace in 1971, and the subsequent publication of the Club of Rome's *Limits to Growth* in 1972. This report employed systems theory and computer models to forecast human development trajectories from 1972 to 2100, warning that exponential growth in resource consumption and pollution emissions would soon exceed the Earth's environmental limits, causing drastic reductions in food production and population. The report sparked a major debate between optimists and pessimists regarding humanity's future, deepening public understanding and recognition of the potential trajectories of human development. *Limits to Growth* is hailed as a milestone with profound, lasting significance in the study of human development issues.

1.2.2 Definition of Sustainable Development

The year 1972 marked a significant milestone in the history of sustainable development. The UN held its first Human Environment Conference in Stockholm and established the UN Environment Program (UNEP). The conference marked the beginning of a new era in global environmental protection efforts. It adopted the Declaration of the UN Conference on the Human Environment, urging governments and people worldwide to collaborate to maintain and improve the human environment for the benefit of all people and future generations. The declaration covered a wide range of topics, including environmental protection, pollution prevention, promotion of economic and social development, and the guarantee of human rights and survival rights in underdeveloped areas, thereby laying the groundwork for future UN declarations on sustainable development.

Beginning in the late 1970s, Western scholars began extensively discussing the concept of sustainability, publishing numerous influential books on the subject. They integrated current advances in the natural sciences and ecological economics to present a vision of a sustainable society and addressed research areas such as environmental issues, climate change, population dynamics, production and consumption patterns, renewable energy utilization, and sustainable urban and agricultural systems.[21] The term sustainable development first appeared in the 1980 report, "World Conservation Strategy," jointly published by the International Union for Conservation of Nature (IUCN), UNEP, and the World Wildlife Fund (WWF).

In 1983, the UN established the World Commission on Environment and Development (WCED), chaired by Norway's Prime Minister Gro Harlem Brundtland. The UN aimed to use this organization to set a transformative agenda for the world, with sustainable development as the fundamental goal. After 4 years of effort, in 1987, the WCED released a report titled, "Our Common Future," also known as the Brundtland Report. The report introduced the widely used definition of sustainable development: "Sustainable development is development that meets the needs of the present without compromising the ability of future generations to meet their own needs." This definition incorporates principles of fairness, sustainability, and commonality, emphasizing two key points: the necessity of

[21] In 1967, American historian Lynn White, Jr. published *The Historical Roots of Our Ecological Crisis*, which reviewed the ecological crisis that emerged in the 1960s. In 1968, American ecologist Garrett Hardin published *The Tragedy of the Commons*, pointing out that limited resources are destined to be overexploited as a result of free use and unlimited demand.

development, especially for the poor, and the inherent limits of development in order to avoid jeopardizing future generations' survival and development.

The Brundtland Report contains three main sections: "Common Concerns," "Common Challenges," and "Common Endeavors." It examines global population, food security, species and genetic resources, energy, industry, and human habitation, systematically discussing major economic, social, and environmental issues facing humanity. The report makes three critical observations: (1) environmental, energy, and development crises are interconnected; (2) the earth's resources and energy are insufficient to meet human development needs; and (3) the development model must change for the benefit of current and future generations. "Our Common Future" notes that the unequal use of "ecological capital" is the "main environmental problem and the main development problem on the planet,"[22] asserting that developed and underdeveloped countries can achieve sustainable development only through joint cooperation to reduce inequality. Thus, sustainable development has three core features: the sustainable use of natural resources and a healthy ecological environment, driven by sustainable economic development, aimed at achieving comprehensive social progress.

The United Nations Conference on Environment and Development (UNCED), held in Rio de Janeiro, Brazil, in 1992, was another major breakthrough in the history of sustainable development. The conference issued the Rio Declaration, which, for the first time, recognized sustainable development as a human right. Reflecting the spirit of the Brundtland Report, the Rio Declaration clarified the conflicting, yet interdependent, relationship between environment and development and outlined the common but differentiated responsibilities of various countries in achieving sustainable development. The conference introduced the well-known Agenda 21, which summarizes our understanding of sustainable development theory and proposes the United Nations Framework Convention on Climate Change (UNFCCC) and the Convention on Biological Diversity (CBD). The UNFCCC was succeeded by the Kyoto Protocol[23] in 1997.[24] Afterward, the concept of sustainable development achieved global consensus.

[22] World Commission on Environment and Development. Our Common Future [M]. Translated by Wang Zhijia, et al. Changchun: Jilin People's Publishing House, 1997:7.

[23] The Kyoto Protocol is a supplementary provision to the United Nations Framework Convention on Climate Change (UNFCCC), established in Kyoto, Japan in December 1997. Its goal is to stabilize the concentration of greenhouse gases in the atmosphere at an appropriate level to ensure the smooth adaptation of ecosystems, the safe production of food, and the sustainable development of the economy. https://unfccc.int/resource/docs/convkp/kpeng.pdf

[24] European Commission. Report of the Commission of the European Community to the United Nations Conference on Environment and Development [R]. Rio de Janeiro, Brazil: United Nations Publication, 1992.

1.2.3 Global Consensus and Current Actions

Starting in the 1990s, discourse on sustainable development expanded beyond the UN and began to permeate research institutions, universities, and even the commercial and financial sectors. Universities in Canada, the UK, and the US started offering courses and degrees in sustainable development, establishing it as an emerging academic field. Additionally, international financial institutions such as the World Bank began incorporating the concept of sustainable development into their reports, emphasizing that a healthy environment is crucial for sustainable development and a robust economy.[25]

One of the most influential concepts in business is the triple bottom line (TBL) that John Elkington proposes. The TBL refers to the economic, environmental, and social bottom lines. Elkington argues that for a company to maintain long-term survival and profitability, it must prioritize environmental quality and social welfare, strive to narrow the wealth gap, and promote health and education. A sustainable company should not profit from human suffering, nor should it exacerbate social problems.[26]

At the beginning of the 21st century, 189 countries adopted the United Nations Millennium Development Goals (MDGs) during the UN Millennium Summit. The MDGs included eight goals to be achieved by 2015: eradicating extreme poverty and hunger, achieving universal primary education, promoting gender equality and empowering women, reducing child mortality, improving maternal health, combating HIV/AIDS and other diseases, ensuring environmental sustainability, and developing a global partnership for development.[27] Reducing global poverty was the core goal of this plan. By 2015, although significant progress had been made, some development goals remained unmet.[28] Notably, however, China played a crucial role in achieving the poverty reduction and other Millennium Development Goals (Fig. 1.4).

As a continuation of this ambitious global initiative, in September 2015, at the United Nations Sustainable Development Summit, 193 member states formally adopted the United Nations Sustainable Development Goals (SDGs). These goals include: No Poverty (the primary goal), Zero Hunger, Good Health and Well-being, Quality Education, Gender Equality, Clean Water and Sanitation, Affordable and Clean Energy, Decent Work and Economic Growth, Industry, Innovation and Infrastructure, Reduced Inequalities, Sustainable Cities and Communities,

---

[25] Pezzey, J. Sustainable Development Concepts: An Economic Analysis [J]. Washington, D.C.: The World Bank, 1992.

[26] Elkington, J. *Cannibals with Forks: The Triple Bottom Line of twenty-first Century Business* [M]. Oxford: Capstone Publishing LTD, 1997.

[27] See the United Nations website https://www.un.org/zh/millenniumgoals/

[28] Xinhua Net Summary: Significant achievements of the United Nations Millennium Development Goals but not fully achieved [EB/OL]. (2015-07-07) [2021-09-26]. http://www.xinhuanet.com/world/2015-07/07/c_1115841125.htm

Fig. 1.4 The 17 Sustainable Development Goals of the United Nations (Picture source: https://www.un.org/sustainabledevelopment)

Responsible Consumption and Production, Climate Action, Life Below Water, Life on Land, Peace, Justice and Strong Institutions, and Partnerships for the Goals, for a total of 17 objectives. Adoption of the SDGs was a call to all countries—whether poor, wealthy, or middle-income—to take collective action and to promote economic prosperity while protecting the planet. The aims were to comprehensively and thoroughly address social, economic, and environmental development issues by 2030 and to ultimately transition to a sustainable development path.[29]

At the time of writing this book, the COVID-19 pandemic has not yet been eradicated, and less than a decade remains until the 2030 deadline. The task of fully achieving the SDGs is arduous and may seem nearly impossible. However, the significance of the SDGs lies in the fact that humanity shares a common story and a beautiful vision for the future, guiding us toward a path of sustainable development.

1.2.4 China's Exploration of Sustainable Development

In the half-century since the global sustainable development movement came into existence, China has actively participated, taking on the responsibilities expected of a populous, economically significant country on the world stage. It has explored

[29] United Nations. Sustainable Development Goals [EB/OL].[2021-09-18]. https://www.un.org/sustainabledevelopment/zh/

local paths toward common prosperity and sustainable development. Beginning from the first Human Environment Conference in 1972 to the present, China's sustainable development efforts have traversed through four distinct stages.[30]

First Stage: Clarifying Environmental Protection Strategy and Promoting Environmental Legislation: In 1972, a year after China officially returned to the UN, it participated in the first Human Environment Conference. In August 1973, the State Council convened the first National Environmental Protection Conference and passed China's first environmental protection document, "Several Provisions on Protecting and Improving the Environment," establishing the environmental protection strategy of "protecting the environment and benefiting the people." In 1978, the constitution included for the first time a mandate that "the state protects the environment and natural resources and prevents pollution and other public hazards." In 1979, the National People's Congress designated March 12 as Tree Planting Day, and China began building the Three-North Shelter Forest. The 1982 Constitution further mandated the rational use of natural resources and the protection of precious animals and plants. In December 1989, the Seventh National People's Congress passed the Environmental Protection Law of the People's Republic of China, marking the first major environmental legislation after the founding of New China.

Second Stage: Establishing the Concept of Sustainable Development: In 1992, China participated in the UN Conference on Environment and Development in Rio de Janeiro and committed to implementing the Agenda 21 document. In March 1994, China issued the China Agenda 21. By 1995, China had officially incorporated the sustainable development strategy into its Ninth Five-Year Plan, emphasizing the need for coordinated and sustainable development of the economy and society. This was the first time the term sustainable development appeared in an official national document. In September 1997, at the Fifteenth National Congress of the Communist Party of China, Comrade Jiang Zemin emphasized the need to implement a sustainable development strategy as part of modernization. In November 2002, at the Sixteenth National Congress, Jiang Zemin included enhancing sustainable development capabilities as a goal in building a moderately prosperous society, thereby solidifying China's sustainable development concept.

Third Stage: The Scientific Development Concept of Ecological Civilization: In October 2003, at the third plenary session of the Sixteenth Central Committee, Comrade Hu Jintao proposed the scientific development concept of "adhering to people-oriented, establishing a comprehensive, coordinated, and sustainable development view, and promoting comprehensive development of the economy, society, and people." At the Central Population, Resources, and Environment Work Symposium in 2005, Hu Jintao introduced the concept of "ecological civilization." The Seventeenth National Congress of the Communist Party of China integrated the construction of ecological civilization into the goal of building a moderately

[30] Qiao Qingju. Construction of Ecological Civilization in China Since the Reform and Opening Up [EB/OL]. (2019-01-04) [2021-09-26]. http://theory.people.com.cn/n1/2019/0114/c40531-30525604.html

prosperous society, emphasizing the need to build a resource-conserving, environmentally friendly society based on natural laws and aimed at sustainable development. The 2010 Twelfth Five-Year Plan further solidified green, low-carbon development concepts, focusing on energy conservation and reductions in emissions, and for the first time, included carbon emission intensity as a binding indicator.

Fourth Stage: The Inclusive Beautiful China Construction Strategy: In 2012, members of the Eighteenth National Congress of the Communist Party of China required the integration of construction of an ecological civilization into all aspects and processes of economic, political, cultural, and social development, incorporating the concept of Beautiful China into the theoretical framework of socialism with Chinese characteristics. The Nineteenth National Congress made "adhering to the harmonious coexistence of man and nature" a fundamental strategy of "Xi Jinping's Thought on Socialism with Chinese Characteristics for a New Era," emphasizing that modernization must include harmony between man and nature. The proceedings highlighted the need to create material and spiritual wealth while providing high-quality ecological products to meet the public's growing needs for a beautiful ecological environment.[31]

China's exploration of sustainable development has been innovative, evolving from early environmental protection efforts to a broad sustainable development concept that includes economic, social, and cultural dimensions. While actively learning from experiences around the globe, China also takes into account its specific development stages and socioeconomic background. It engages in continuous enrichment and expansion of sustainable development's connotations and thus contributes to building a "community of shared future for mankind," promoting global ecological security and sustainable development for humankind.

1.3 Core Issues of Sustainable Development

The course of human development is a process of gradually breaking away from the constraints of nature and developing human civilization. Historically, we have lacked sufficient awe and respect for nature, engaging in blind exploitation and plundering, which has led to increasingly serious ecological disasters. Additionally, persistent conflicts among humans, characterized by the strong bullying the weak, have resulted in social injustice. Only when various crises emerged did humans begin to realize that the limitations of nature serve as a form of protection, making us realize that humans and nature are interconnected, and that protecting nature is synonymous with protecting humanity. To create a hopeful future, we must

[31] Xi Jinping. Winning the comprehensive construction of a moderately prosperous society and achieving the great victory of socialism with Chinese characteristics in the new era—Report at the 19th National Congress of the Communist Party of China [R/OL]. (2017-10-27)[2021-09-16]. http://www.gov.cn/zhuanti/2017-10-27/content_5234876.htm

prioritize the coordinated development of the environment, society, and economy. Without this balance, sustainable development remains a mere dream and political slogan.

1.3.1 Environmental Dimension of Sustainable Development

Humans depend on various elements for survival that the natural environment provides. Throughout history, in every social configuration we have assumed, human impact on the natural environment has assumed two primary forms: resource acquisition and waste discharge. *From an ecosystem perspective, humans obtain energy and materials from nature (input) and simultaneously discharge transformed waste gases, wastewater, and waste into nature (output).* This perspective allows us to understand the sustainability of the environment.

In terms of input, humanity faces an increasingly severe resource crisis. The Millennium Ecosystem Assessment the UN published in 2005 points out that humans have consumed two-thirds of the Earth's natural resources.[32] The World Wildlife Fund's 2006 Living Planet Report stated that, if resource consumption continues at the current rate, by 2050, humans will have consumed the equivalent of the natural resources of two Earths. The term sustainability emerged in relation to the energy crisis in Europe at the time.[33] Food, energy, and land are the basic resources we need as humans for survival. Due to rapid population growth, humans are continuously expanding cultivation areas, reclaiming swamps, forests, and grasslands to grow crops and domesticate animals, leading to the rapid disappearance of natural habitats. The Millennium Ecosystem Assessment noted that, in the past 60 years, the area of land claimed globally exceeds the total claimed in the 18th and 19th centuries, with humans having acquired 24% of the Earth's land surface for cultivation.[34] Simultaneously, with rapid urbanization, human settlements have evolved from villages to small towns and from small towns to cities. The expansion of the man-made environment continually shrinks natural spaces, making occasional incursions by wild animals into cities notable news.

---

[32] The Millennium Ecosystem Assessment is a report the United Nations completed that involved 95 countries and 1360 experts over 4 years at a cost of $24 million. The conclusions, predictions, and recommendations of the report have significant implications for future ecological conservation.

[33] As mentioned above, the term sustainable first appeared in Carlowitz's *Forestry Economics*. The father of American ecological ethics and the author of *A Sand County Almanac* (1949), Aldo Leopold, also originated in forestry. Well before the end of the nineteenth century, wood, as the world's main fuel, had been in short supply since the Middle Ages in Europe. For example, Portuguese ships used for great voyages had to be built in colonies like India; the raw materials for the Spanish Armada had to be purchased from Poland.

[34] Zhao Tudong. Millennium Ecosystem Assessment Report Collection [M]. Beijing: China Environmental Science Press, 2007.

Human activities have expanded beyond the necessities for survival to unprecedented levels in the pursuit of development. The extraction of coal, oil, metals, and rare minerals require large-scale cooperation and advanced technical equipment and entail continuously digging below the surface, into the ocean, and within mountains. The greatest negative externality of high-intensity resource development is the fundamental destruction of habitats. Many animals and plants lose their habitats, their numbers decline, their populations become extinct, leading to a loss of biodiversity and disrupting the balance of the ecosystem. This trend suggests that future generations might inherit a planet that is resource-poor, species-poor, and lacking vitality.

In terms of output, the large amount of waste human production and consumption generates, including polluted air, wastewater, and solid waste, has had significant environmental impact, such as global warming, a decrease in the ozone layer, smog, acidification, and eutrophication. Among these, global warming poses the most serious threat. The first working group IPCC Sixth Assessment Report released in August 2021 stated that *human activities have caused unprecedented climate warming*. The report noted that the current global surface average temperature is about 1°C higher than pre-industrial levels, with projections indicating that the global temperature increase will reach or exceed 1.5°C in the next 20 years. Considering all emission scenarios, the global surface temperature will continue to rise at least until the mid-21st century, potentially exceeding 1.5°C or even 2°C. Continued global warming is expected to intensify the global water cycle, increasing variability, monsoon precipitation, and the intensity of dry and wet events.[35] As a result, some regions will endure extreme rainfalls and floods, while others will experience more severe droughts than previously.

Besides the catastrophic climate changes global warming causes, environmental pollution directly affects human health. Incidents such as the London smog even, the Chornobyl nuclear disaster, and Minamata disease in Japan have had long-term effects on the health of tens of thousands of people, especially those who are underprivileged and unable to escape polluted areas. Even more concerning is the long-term damage environmental pollution inflicts on the biosphere. Polluted water bodies and soil often take a very long time to recover.

For hundreds of years, human activities have had a significant negative impact on natural resources and the environment, gradually affecting all the resources, air, soil, and water upon which all organisms rely for survival. The populations of some vulnerable species have decreased or become extinct, and the once-balanced ecosystem has been damaged, ultimately threatening the continuation and development of human populations. In the following chapters, we will discuss in detail various design strategies to reduce environmental impact. Overall, these strategies include reducing the material and energy intensity of human activities in terms of input, conserving increasingly scarce resources, utilizing renewable resources, recycling and reusing materials, and designing non-material product-service systems. In terms of output, our proposals include reducing waste generation and emissions,

[35] IPCC. Climate change 2021: the physical science basis [m]. Cambridge: Cambridge University Press, 2021.

minimizing pollution of air, water, and soil, such as through the development and innovative design of biodegradable materials and technologies, and the use of clean energy. However, it is important to note that all these efforts depend on reshaping prevailing production modes and lifestyles, rather than implementing gradual improvements.

1.3.2 Sustainable Development: The Social Dimension

Social sustainability garnered attention later than environmental sustainability and to date lacks universally accepted definitions. Social sustainability, in the broadest sense of the term, refers to the continuation and long-term positive development of human society. Within the context and goals of sustainable development, social sustainability focuses on the equity principle, which asserts that every person on the planet has the right to the space and resources necessary for survival and development. If we align a broad understanding of social morality and ethics with the goals of sustainable development, it enriches the connotations of social sustainability: "Creating equal opportunities based on respecting basic rights and cultural diversity; promoting the establishment of a democratic, inclusive, united, healthy, safe, and fair society, while opposing all forms of discrimination."[36] Having such a comprehensive definition helps recognize the importance of evaluating the sustainability of the social.

The United Nations Development Program (UNDP) adopted the Human Development Index (HDI) in its 1990 Human Development Report to measure the socioeconomic development levels of UN member countries, aiming to correct the previous reliance on a single Gross Domestic Product (GDP) metric for assessing development progress. The HDI comprises three indicators: expected lifespan, adult literacy rate, and per capita GDP, to reflect the level of social development through human longevity, knowledge, and living standards. While social development is often predicated on high-level economic development, economic prosperity alone does not necessarily bring about social equity and progress. Social development is typically determined by a society's internal resource and wealth distribution model.

From the perspective of sustainable development, the HDI links social development with economic development but lacks consideration of ecological costs. Therefore, some scholars propose a more precise definition of social development as "*achieving a higher level of welfare within the carrying capacity of the ecological environment*." This definition directly connects the goals and costs of social and economic development, indicating the direction and path for sustainable development.

---

[36] PRALL U. "The Sustainability Strategy of the European Union [J]." *Journal for European Environmental & Planning Law*, 2006, 3(4): 325–339.

Historically, evaluations and comparisons of different development models have often relied on the achievements of economic and social development achievements. However, when we include ecological costs in assessments, it becomes evident that *if the current development model continues, neither rich nor poor countries will achieve sustainable development.* Vaclav Smil, author of *Energy and Civilization: A History*, posits that at low stages of development, the use of energy (ecological cost input) has a linear relationship to the HDI. However, at advanced stages, increased energy input does not correspond to increased social development. For example, despite similar levels of development, the United States consumes nearly twice as much energy as Germany and France. Clearly, such a socioeconomic model is unsustainable (Fig. 1.5).

Some scholars have proposed four models of social development using the two dimensions of the Human Development Index (HDI) and ecological footprint.[37] These models define four quadrants according to welfare levels (measured by the HDI) and natural consumption (measured by the ecological footprint): "high welfare, low consumption" (the only sustainable development model), "high welfare, high consumption," "low welfare, high consumption," and "low welfare, low consumption." Researchers used HDI data the UN Development Program published in 2015 and global ecological footprint data to analyze 82 countries with populations of over 10 million. The results show that no country is currently in the "high welfare, low consumption" quadrant. Nineteen countries are in the "high welfare, high consumption" quadrant, of which 17 are developed countries, and two (Argentina and Chile) are developing countries. Twenty-five countries are in the "low welfare, high consumption" quadrant, including 23 developing countries and two developed countries (Russia and Romania). Forty-eight countries fall into the "low welfare, low consumption" quadrant, with 14 being developing countries and 24 being the least developed countries. In the "high welfare, high consumption" quadrant, the ecological footprints of Australia, the United States, and Canada are 5.37, 4.75, and 4.72 times their ecological carrying capacity, respectively. This means that if all global residents lived like them, at least 4–5 Earths would be needed to meet human natural consumption needs (Fig. 1.6).

The UN emphasizes two aspects of fairness in its definition of sustainable development. First, as trustees, we have an obligation to protect natural and cultural resources for future generations, thereby achieving "intergenerational equity."[38] Second, all people within the current generation, regardless of nationality, race,

[37] Zhang Shuai, Shi Qinghua. Research on Sustainable Development Applying Human Development Index and Ecological Footprint—Based on the Strong Sustainability Research Paradigm[J]. Journal of Shanghai Jiaotong University: Philosophy and Social Sciences Edition, 2017, 25(3): 99–108.

[38] Intergenerational equity is an important aspect of the principle of sustainable development, mainly referring to the current generation's demand to preserve naturalgeneration of humans is the trustee of future generations. Under this trusteeship, the current generation has the responsibility to protect the earth's environment and hand it over to future generations intact. Reference: Baier A C. For the sake of future generation[m]//regan t. Earthbound: new introductory essays in environmental ethics. New York:John Wiley & Sons, 1984: 214–215.

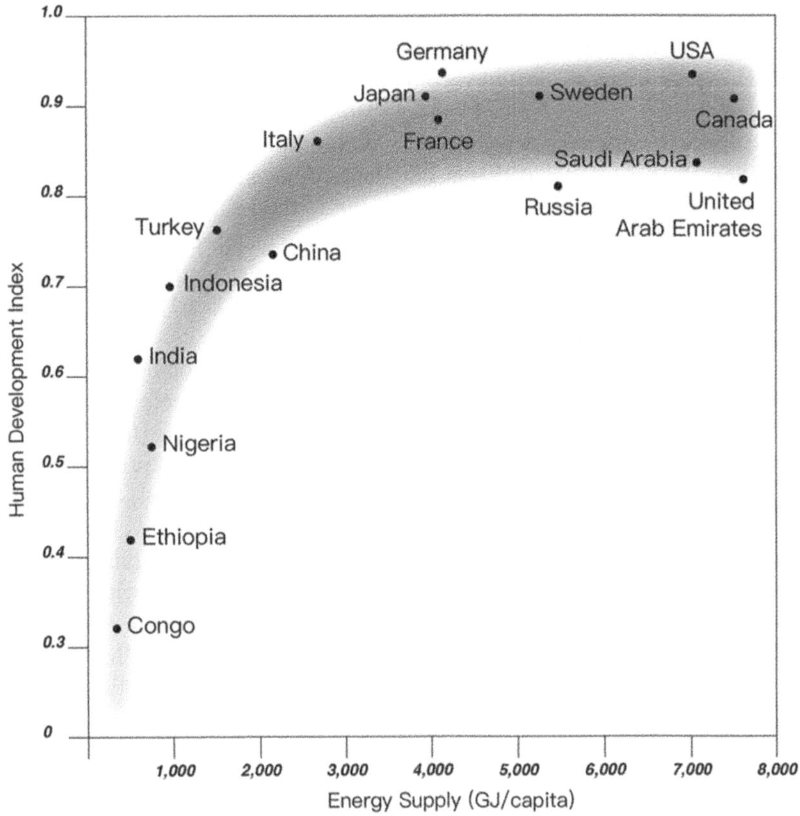

Fig. 1.5 Smil. Energy and Civilization [M]. Translated by Wu Lingling, Li Zhu. Beijing: Jiuzhou Press, 2021:367. Original data from the United Nations Development Programme (2015 year) and the World Bank (2015 year). Vaclav Smil, Energy and Civilization: A History

gender, economic development level, or culture, should have equal rights to use public natural resources and enjoy a clean, healthy environment. This implies that the development of some people today cannot come at the expense of the development of others. Since developed countries have achieved a "high welfare" society through "high consumption," whether "low welfare" countries can improve their overall welfare through the rational use of natural resources will be a crucial issue in future international politics. Additionally, within a country, achieving the fair possession and distribution of resources among different groups is a prerequisite for attaining social sustainability.

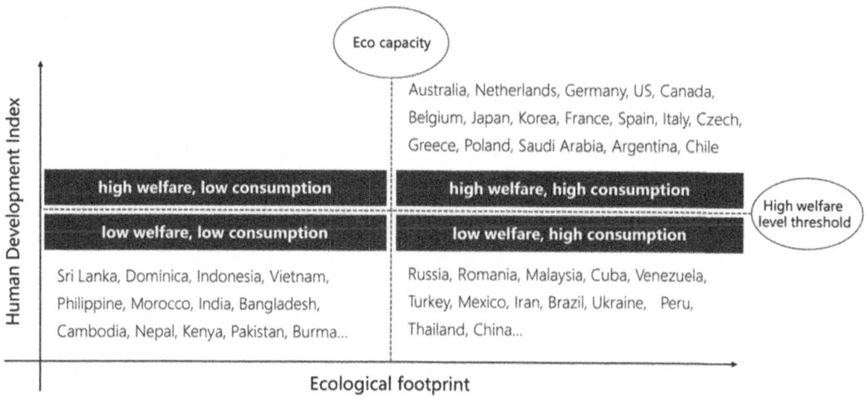

Fig. 1.6 Sustainable development quadrant based on the Human Development Index and ecological footprint

1.3.3 The Economic Dimension of Sustainable Development

Economic sustainability is inseparable from environmental and social sustainability. In fact, purely economic sustainability does not exist. The terms economy and ecology share the same etymology, underscoring the intrinsic link between economic behavior and natural ecology. In classical economics, the focus is on the allocation of finite or scarce resources among competing purposes.[39] Within the framework of sustainability, an important topic of recent theoretical research is how to ensure that economic development serves sustainable development, rather than assuming that economic growth must come at the expense of the environment.

In classic economics, the natural environment's primary function is to provide the resources needed for human survival and development. In this understanding, these resources are competitive, exclusive goods, priced and traded in the market. However, the advent of ecological economics introduced a systemic view of the natural environment, which reveals that natural resources cannot be separated or priced independently of their overall ecosystem. For instance, air and oceans are indivisible and cannot be exclusively owned. The felling and selling of trees may cause soil erosion and biodiversity loss and reduce air purification, impacts not reflected in the price of timber but borne by society. Similarly, when enterprises emit waste gases and wastewater, the benefits accrue to the enterprise, while the environmental damage is a social cost.

[39] Daly, Farley. Ecological Economics: Principles and Applications[M]. Translated by Jin ZhiNong, et al. Beijing: Renmin University of China Press, 2014: 3.

Therefore, in the economic dimension of sustainable development, cost and benefit need redefinition. In 1923, John Mill first proposed the concept of a steady state in his book *Principles of Political Economy*. By 1973, economist Herman Daly edited *Toward a Steady-State Economy*, thus establishing the notion of steady-state economics. This latter branch of economics critiques modern economics for focusing on the allocation of scarce production factors and income distribution, while assuming indefinite high-speed growth, ignoring the fact that rapid growth depletes raw materials, exacerbates pollution, and undermines the material basis for growth.

Following steady-state economics, the concept of degrowth emerged in Western countries. Coined by French social philosopher André Gorz in 1972, degrowth gained wider recognition at a 2008 conference. In 2016, the Club of Rome published *Reinventing Prosperity*,[40] which suggests that the ecosystem can no longer support rapid growth and that a 1% growth rate is sufficient.

Both steady-state economics and degrowth respond to sustainability as needed in Western contexts. However, in a world with stark socioeconomic disparities, where many still struggle with basic needs, the absence of capital and technology transfer from rich to poor countries and the lack of extensive international cooperation make steady state and degrowth impractical. Only by acknowledging that all individuals and societies have the right to aspire to living a good life, and by accommodating differences in systems and cultures, can we reach consensus and explore a sustainable economic paradigm.

In sustainable development as theoretical system, the environment, society, and the economy are interdependent, mutually restrictive, and inclusive. Human survival and development depend on and are constrained by the natural environment. Economic development enables humans to utilize nature's resources effectively and potentially overcome natural constraints. However, if there is no consensus and cooperation on the use and distribution of natural resources—a shared public resource of humanity—intense social conflicts will arise. Given population growth and resource scarcity, an unsustainable society will inevitably lead to environmental and economic unsustainability. Therefore, achieving human sustainable development requires systematic thinking and integration of environmental, social, and economic dimensions

1.4 Theoretical Controversy

First proposed as an idea in the 1980s, in the 1990s, the UN established sustainable development as a global development strategy that has evolved over more than 30 years. Although most researchers agree on the main problems facing humanity

---

[40] Wei Bole, Vijaykumar. Flipping the Limits: The Awakening Path of Ecological Civilization[M]. Translated by Cheng Yiheng. Shanghai: Tongji University Press, 2019.

today and the goals of sustainable development, there are still many controversies, and even conspiracy theories about issues such as environmental crises and climate change. These controversies are mainly about varying models of strong and weak sustainability, anthropocentrism, technocentrism, and ecocentrism.

1.4.1 Paradigm of Sustainable Development: Weak Sustainability and Strong Sustainability

Western economists refer to the two basic economic functions of nature—namely, source function and self-purification function—collectively as natural capital. Natural capital comprises the stock of natural resources and environmental assets that can provide valuable product or service flows now or in the future.[41] In sustainability economics, human capital and production capital are often collectively termed manufactured capital to distinguish them from natural capital.

Since sustainable development entered academic discourse, economists have developed two distinct paradigms with regard to theoretical and policy considerations.[42] One approach, rooted in mainstream neoclassical economics, operates within the paradigm of economic growth research. The weak sustainability model, as it is called, does not acknowledge that there are natural limits to economic growth.[43] Nobel laureate in economics Robert Solow argued that sustainability involves distributive justice, specifically the equitable sharing of welfare between current and future generations. According to Solow, preserving natural capital is unnecessary as long as sufficient man-made capital is left for future generations to meet their needs.[44]

The weak sustainability model posits that the sum of man-made and natural capital is what matters, not natural capital per se. Therefore, it asserts that current generations can deplete non-renewable resources or emit carbon dioxide, provided they produce enough man-made infrastructure as compensation.[45] Neoclassical economics extends the concept of capital to the natural realm, in line with Ricardo's assumption of the relative scarcity of resources. It effectively treats nature as a tool and

---

41 Hawken, Amory Lovins, Hunter Lovins. Natural Capitalism: The Next Industrial Revolution[M]. Translated by Wang Naili, Zhu Dajian, Gong Yitai. Shanghai: Shanghai Popular Science Press, 2000: 7. Paul Hawken, Amory Lovins, L. Hunter Lovins, Natural Capitalism: Creating the Next Industrial Revolution

42 Pearce d W, et al. Blueprint 2: greening the world economy [m]. London: Routledge, 1991.

43 Daly H E. Georgescu-Roegen Versus Solow/STIGLITZ [J]. Ecological Economics, 1997, 22(3): 261–266.

44 Solow. Economic Growth Theory: An Exposition [M]. 2nd Edition. Translated by Zhu Baohua. Shanghai: Gezhi Publishing House, Shanghai People's Publishing House, 2015.

45 Neumayer. Strong and Weak: Two Opposing Sustainability Paradigms [M]. Translated by Wang Yintong. Shanghai: Shanghai Translation Publishing House, 2006: 1. Eric Neumayer, Weak versus Strong Sustainability: Exploring the Limits of Two Opposing Paradigms

assuming that man-made capital is scarce while natural capital is abundant. While this approach seemed to make sense during the industrial revolution, it has led to a global shortage of natural capital and its services, exacerbated by uneven distribution between developing and developed countries. Moreover, this assumption has become increasingly problematic as economic growth has alleviated the scarcity of man-made capital but intensified the shortage of natural capital.[46] This is what Herman Daly, the main theorist of sustainable development economics, highlights as a transition from an empty world rich in natural capital to a full world constrained by ecological limits, which makes the inadequacies of the weak sustainability paradigm evident.

In contrast, the strong sustainability model contends that human development since the 1970s necessitates recognizing natural limits to economic growth and reforming neoclassical economic theories and methods. This approach, embedded in the relationship between ecosystems and economic systems, addresses the shift from economic growth to improving the welfare of populations. International research in sustainability economics and sustainability science employs this model.

The strong sustainability model argues that all types of capital are essential production factors and that they are complementary and not entirely mutually substitutable. *This paradigm calls for maintaining at the necessary level each subsystem, species, and physical stock of each type of capital in the natural environment. To achieve sustainable development, the stock of each type of capital must be above a reasonable lower limit.*[47] Advances in ecology and complexity theory make it increasingly clear that the intricate relationships within natural systems are often immeasurable and not monetizable. The environment ultimately constrains human society and economic development. This model dominates sustainable development theory.

1.4.2 Anthropocentrism

In 2000, Nobel laureate in Chemistry, Paul Jozef Crutzen, introduced the concept of the Anthropocene to highlight the significant role humans now play in geology and ecology. Crutzen argues that, since the Industrial Revolution in the late 18th century, human interactions with nature have intensified to the point where humans have become a major force influencing environmental changes. The impacts of human activities on Earth's system will persist far into the future, possibly for 50,000 years,

[46] Zhu Dajian. Beyond Growth: How Sustainable Development Economics Differs from Neoclassical Economics [J]. Academic Monthly, 2013(10): 79–89.

[47] Neumayer. Strong and Weak: Two Opposing Sustainability Paradigms [M]. Translated by Wang Yintong. Shanghai: Shanghai Translation Publishing House, 2006: 30.

making it essential to study the Earth from the perspective of the Anthropocene to understand the profound and lasting effects humans have on the environment.[48]

Despite ongoing debates about the precise definition of the Anthropocene, its influence extends beyond geological science. Increasing evidence that human activities directly affect the environment has led many scholars to critically examine the role of anthropocentrism in Western culture. As an ethical framework for the human-nature relationship, anthropocentrism has two main roots in the West. Philosophically, it traces back to the ancient Greek philosopher Protagoras, who posited that man is the measure of all things. Religiously, it is reflected in the Judeo-Christian tradition, particularly in the Book of Genesis, in which God creates humans and commands them to steward other creatures. These philosophical and religious views fundamentally shape Western culture.

The European Renaissance and the Enlightenment further promoted anthropocentrism. French philosopher Descartes established a reductionist epistemology, that viewed animals as machines and suggested that humans have no direct obligations to nature or animals except as these affect humans themselves. Similarly, Kant argued that reason endows humans with intrinsic value, whereas animals and plants, lacking reason, have no moral status. He wrote, "so far as animals are concerned, we have no direct duties. Animals are not self-conscious and are there merely as a means to an end. That end is man… Our duties toward animals are merely indirect duties toward humanity."[49]

However, the environmental movement of the 1960s led many Western scholars to see anthropocentrism, which treats nature instrumentally, as a primary cause of ecological degradation. To counter this, they advocated for anti-anthropocentrism, a perspective that challenges the exploitative relationship humans have with nature.

However, French philosopher Bruno Latour critiques both anthropocentrism and anti-anthropocentrism as dualistic claims that oppose nature and humans.[50] He argues that while this dualistic thinking has propelled modern science and technology, it also harbors significant flaws. Modern complex science acknowledges the interconnectedness of nature, recognizing that humans, despite their significant impact, are part of nature and not separate from it. Thus, both anthropocentrism and anti-anthropocentrism misinterpret the essence of nature.

All the same, questioning anti-anthropocentrism does not justify the exploitation of nature. The unrestrained destruction and plundering of natural resources, which treats nature as a mere object of human conquest, is short-sighted instrumentalism. Protecting the environment and understanding nature ultimately serve to promote the long-term survival of the human species, aligning with a durable,

[48] Christian. *Origin Story: A Big History of Everything* [M]. Translated by Sun Yue. Beijing: CITIC Publishing Group, 2019: 350.

[49] Kant, I. Grundlegung zur Metaphysik der Sitten; Gregor, M., Timmermann, J., Eds.; Groundwork of the Metaphysics of Morals, Translator; Cambridge University Press: Cambridge, UK, 2015.

[50] Bruno Latour. Politics of Nature [M]. Translated by Mai Yongxiong. Zhengzhou: Henan University Press, 2016: 159.

human-centered development stance. In line with this, Marxist ecological theory advocates for *weak anthropocentrism, positing that the continuation of human society depends on a sustainable biosphere.*[51] *Therefore, reducing the negative impacts of human activities on the biosphere is crucial to sustainable development.*

The dualistic opposition inherent in anthropocentrism is a foundation of Western civilization. However, many non-Western civilizations, such as Chinese civilization, maintain a holistic worldview exemplified by the concept of the unity of man and nature, a perspective that has persisted for over two millennia. Human consciousness shapes subjectivity. However, understandings of the world evolve, marked by changes in human limitations at various stages. Recognizing these limitations and refraining from viewing ourselves as the masters of nature can obviate making the arrogant assumption that nature is an inexhaustible resource.

1.4.3 Technocentrism

Over the past few centuries, scientific and technological developments have liberated productivity, bringing wealth and modern civilization to humanity and helping us navigate various crises. For example, the invention of the steam engine significantly improved coal mining efficiency, addressing the potential energy crisis caused by the depletion of forests and ushering in the machine age. Similarly, the maturation of oil extraction technology in the late 19th century increased the efficiency of energy sources, while the development of renewable energy technologies has alleviated the pressure of the oil crisis on the global economy, offering hope for completely eliminating air pollution.

A fundamental assumption of technocentrism is that technology has its own independent purpose: "it is completely independent, self-regulating, and even legislates for everything… Technocentrism also and still belongs to a variant of anthropocentrism, because its essence is to master and possess nature."[52] Technocentrists are confident that if environmental sustainability is set as the next goal, breakthrough technologies and materials will provide new solutions for humanity. In other words, technocentrism is optimistic, acknowledging the severity of various environmental issues, while firmly positing that technology can solve these problems and achieve unlimited growth within the current social framework.

However, in the current global socioeconomic structure, technology is not the fundamental solution to environmental problems. First, technology is not neutral.[53] Since the technological revolution, substantial capital investment has fueled the

---

[51] Pepper. Eco-Socialism: From Deep Ecology to Social Justice [M]. Translated by Liu Ying. Jinan: Shandong University Press, 2005.

[52] Stiegler. Technics and Time: The Fault of Epimetheus [M]. Translated by Pei Cheng. Nanjing: Yilin Press, 2012: 103.

[53] Pepper. Modern Environmentalism: An Introduction [M]. Translated by Song Yubo, Zhu Danqiong. Shanghai: Shanghai People's Publishing House, Gezhi Publishing House, 2020: 105. David Pepper, Modern Environmentalism: An Introduction

invention, application, and iteration of technology, making the primary goal of technology the realization of capital appreciation. Corporations that monopolize technology, as well as developed countries, are unlikely to voluntarily share their technological advances with other nations. Even in the 21st century, during a global pandemic, achieving technology-sharing for virus detection and vaccine production proved difficult. Additionally, human values and political factors influence technology. Technology is a means, and those who control it—whether individuals or interest groups—determine its use, accessibility, and timing. Consequently, not all technological achievements benefit the public or promote economic and social welfare.

Secondly, there are limitations to technology's ability to deal with complex systems. In 1971, American ecologist Barry Commoner published *The Closing Circle*, in which he argues that technology has relativity, non-association, and controllability. Commoner notes:

The fragmented design of technology is a reflection of its scientific basis. Because science is divided into disciplines, these disciplines are largely dominated by such a concept, that is, it is believed that complex systems can only be understood when they are first broken down into parts. The prejudice of reductionists also tends to hinder basic science from considering real-life problems, such as environmental degradation.[54]

Reality has repeatedly confirmed Commoner's view. For instance, the Green Revolution increased agricultural output, significantly boosting population growth, but the overuse of pesticides and fertilizers degraded soil and water sources, leading to genetically modified technology which gave rise to social controversies and unknown environmental impacts. While technology can be a solution, it also introduces significant risks and uncertainties.

The following story about rice cultivation in Bali illustrates the need for caution in the application of technology:

People on the island of Bali have cultivated rice since the third century BC. Relying on a complex irrigation system, they grow rice in terraced fields on hills and in paddy fields in the lowlands. River water generally flows from high mountains down deep valleys to downstream areas, so in many places, river water needs to be channeled through man-made channels to ditches, and then into terraced fields. After irrigating the paddy fields, the excess water is returned to the river, or slowly flows to lower-lying depressions. Since the tenth century AD, farmers have spontaneously formed basic cooperative units called Subaks, responsible for managing this system, and irrigating or draining according to natural cycles, maintaining good soil fertility for thousands of years, with almost no reduction in yield or crop failure. In conjunction with this rice cultivation technology, Bali has developed a religious system, with religious ceremonies perfectly aligned with the rice cultivation cycle. If there is a disagreement over water distribution between Subaks, it is often coordinated through local temples.

---

[54] Commoner. The Closing Circle: Nature, Man, and Technology [M]. Translated by Hou Wenwei. Changchun: Jilin People's Publishing House, 1997:154.

The population of Indonesia exploded in the 1950s. In 1967, Bali, as a high-yield grain area, imported about 10,000 tons of grain annually. To increase grain production, Indonesia began to learn from the experience of the Mexican Green Revolution and introduced new types of rice; by 1977, 70% of the terraced fields in Bali had been converted to high-yield rice; in 1979, with the funding of the Asian Development Bank, the irrigation system was rebuilt through the Bali Irrigation Plan; in the early 1980s, a new agricultural system centered on high-yield rice varieties and cultivation techniques was established, and the traditional Subak system was completely destroyed.

But the technology that brought high yields also brought problems. A single variety of rice is more susceptible to diseases and pests. For example, IR-8 rice is susceptible to brown planthoppers, which are not usually numerous, but can reproduce rapidly in a monoculture environment. So a new super variety of rice, IR-26, which can resist pests, replaced IR-8. But a new type of planthopper broke through IR-26 and people had to switch to IR-36. Unfortunately, IR-36 is easily infected with the tungro virus, so it was replaced by PB-50. In 1982, PB-50 was infected with a fungus, which destroyed 6000 hectares of rice. Traditional agriculture controlled these natural diseases well through allowing fields to lie fallow, crop rotation, and other techniques, but the Green Revolution abandoned these complex systems, aiming for single high yields. Worse still, the high yields required the heavy use of pesticides and fertilizers, which polluted local water systems. In the mid-1970s, after a brief period of high rice yields, Bali began to import rice again. In 1987, anthropologist Stephen Lansing and systems ecologist James Kremer developed computer models of Bali's two water systems. After comparing the efficiency of various irrigation patterns, they found that the traditional Subak system, coordinated by water temples, produced more, had fewer pests and diseases, and made more efficient use of water than the new system.[55]

1.4.4 Ecocentrism and Deep Ecology

In stark contrast to technocentrism, in contemporary ecological philosophy, anti-anthropocentric thoughts, such as ecocentrism and biocentrism, have emerged. Among these, deep ecology, which renowned Norwegian philosopher Arne Naess espouses, is one of the most revolutionary and challenging ecological philosophies in contemporary Western environmentalism. It also serves as a pioneering environmental value concept.

According to Naess, shallow ecology is anthropocentric, focusing solely on human interests, whereas deep ecology is non-anthropocentric and holistic, concerned with the welfare of the entire natural world. Shallow ecology addresses

[55] Adapted from: Radkau. Nature and Power: A World Environmental History [M]. Translated by Wang Guoyu, et al. Baoding: Hebei University Press, 2004.

symptoms of environmental degradation, such as pollution and resource depletion. In contrast, deep ecology questions the root causes of environmental crises, including social, cultural, and human factors. The natural value theory of deep ecology emerges from contemporary ecological science, which posits that nature is an interdependent community where even the simplest life forms, such as paramecia, play crucial roles in stabilizing the entire biological community. Human life depends on the dynamic balance of the entire ecosystem.

Deep ecology holds that natural objects possess intrinsic value and are not merely tools for human use. It asserts that all biological species are equal, encompassing not only natural species and communities but also all human ethnic groups. Deep ecology opposes waste, advocates for population control, and promotes appropriate rather than dominating technology. People who ascribe to its philosophy seek self-realization through full emersion in both material and spiritual life.

For Western societies, which have propagated anthropocentric views for over two thousand years, deep ecology constitutes a continuation of the thinking of environmentalists such as Leopold, Mumford, and Schumacher. Its adherents directly address the current environmental crisis and attempt to find fundamental solutions. Deep ecology represents a philosophical approach that breaks significantly from mainstream Western intellectual traditions and has a substantial social impact.

Despite this, deep ecology contains inherent contradictions and is strongly utopian. It raises questions we cannot answer. For instance, if natural objects have intrinsic value but cannot assert this value, how should humans interact with them? Humans define the non-economic value of nature, including social, cultural, and aesthetic values, based on higher needs. From nature's perspective, deserts, oceans, mountains, and abysses are merely objective things, neither beautiful nor ugly, neither good nor evil. Value judgments and aesthetic judgments originate from human perspectives. Thus, at its root, deep ecology still harbors anthropocentric elements, albeit with a long-term view aimed at ensuring the continuation of the human species. Even if humans were to forsake development goals and regard survival as their bottom-line requirement, they would still depend on the existence of producers and decomposers in the biosphere. Therefore, any non-anthropocentric viewpoint lacks a factual basis. Deep ecology is romantic, even idealistic, some of its theoretical basis is solid, but its logic is fragmented, its advocacy is sincere, but also powerless.

1.5 How Can Humanity Be Sustainable

Since their appearance on this planet, humans have relied on our superior intelligence and culture to outpace other species, constantly proliferating and growing until we have become the most dominant species in the biosphere. Even after attaining overwhelming advantages and unprecedented "civilized" achievements, humans did not slow our pace of conquering and transforming nature. Instead, our conquests fostered ever stronger desires for possession and control. It is only when we face significant crises in ecological, social, ethical, and economic development that we begin to consider seriously strategies to ensure our future survival.

Is sustainable development possible for humans of the future? Scientists in the 1970s issued warnings that we were approaching the limits to growth. Indeed, today we have already crossed this limit. Various signs indicate that the ecosystem upon which we rely for survival is on the verge of collapse. The World Business Council for Sustainable Development (WBCSD) mentioned in its 1996 Eco-Efficient Leadership Report that, considering foreseeable population growth and the ever-increasing demand for welfare under the current unfavorable environment, only by increasing the ecological efficiency of production and consumption systems by ten-fold can we achieve the requirements for sustainability. In order to create a sustainable system, we need to reduce by 90% the environmental resources we use to provide services from that mature industrial societies currently consume.[56] Although this is only an estimate and requires additional scientific research, it still provides a general parameter what how much we need to change. In the face of such a reality, gradual technological improvements and end-of-pipe governance are clearly inadequate. *Only by turning to source intervention and systemically reshaping current production and consumption methods can we achieve a sustainable social transformation.* Although different countries choose different paths, forward-looking social experiments are also indispensable. However, the ultimate realization of sustainable transformation will definitely rely on a common vision, cooperation, and collective action by humanity.

If the overall relationship between humans and nature is one of environmental sustainability, then the relationships between various groups determines how people share limited environmental resources and bear environmental responsibilities. *If there is not a benign social relationship among groups of people, then environmental problems will succumb to the tragedy of the commons.*[57] *Therefore, to some extent, social sustainability is the prerequisite for environmental sustainability.* In addition to resource and environmental issues, scientists often cite population growth as a reason for unsustainability. However, history shows that only when social and economic development reaches a certain level is there a voluntary decline in population. High fertility rates are often a compensatory strategy for high mortal-

---

[56] WBCSD. Eco-Efficient Leadership For Improved Economic And Environmental Performance[R/OL]. 1996 [2021-09-26]. http://wbcsdservers.org/wbcsdpublications/cd_files/datas/wbcsd/business_role/pdf/ee leadershipforimprovedeconomic&enviperformance.pdf.

[57] In 1968, American scholar Garrett James Hardin published "The Tragedy of the Commons" in the journal *Science*. The article outlines a scenario where a group of herdsmen graze together on a common pasture. One herdsman wants to raise one more sheep to increase his personal gain, even though he knows that there are already too many sheep on the pasture and increasing the number of sheep will degrade the quality of the pasture. How will the herdsman make a choice? If everyone acts out of self-interest, they will definitely choose to raise more sheep for more profit because the cost of pasture degradation is borne by everyone. The tragedy of the commons is often used to describe the difficulty in allocating non-exclusive, competitive public resources.

ity rates and short life expectancies.[58] In fact, whether we are speaking of rich or poor countries, or rich people or poor people, in modern society, dominant groups cannot guarantee their ability to take care of themselves. Only by viewing the global village as a whole and upholding the concept of a community of shared human destiny can we jointly commit to the protection of the biosphere. Such a collective commitment provides hope for the sustainable development of human society.

[58] Sun Wenzhong. New Theory on Population Transition—On the Theory of Population Quantity and Quality Development [J]. Population and Economy, 2008(S1):7.

Chapter 2
Design for Sustainability

Abstract Design for Sustainability (DfS) originates from the concept of sustainable development. It represents the design industry's theoretical reflection on the relationship between economic development and the environment and society, and design practices that constantly seek change and transformation based on this relationship. This conceptual evolution has shifted from a relatively isolated, technology-centered perspective to a more systematic, human-centered approach. Simply put, DfS is the thinking and practice of design researchers and practitioners within a broad context of sustainable development.

Keywords Green design · Ecological design · Product-service system design · Social design · Sustainable transformation

The concept of DfS is not closed. On the one hand, it bears a close relation to concepts such as green design, ecological design, and social design. On the other, it has its own characteristics and systematic approach. Over time, DfS has absorbed decades' worth of innovative ideas and theoretical methods from resource management, environmental and ecosystem studies, research on social change, and business models, enriching its content and making it multidimensional. *Essentially, any design teaching, practice, and research activities that practice sustainable development fall within the scope of DfS.*[1]

Although humans have long employed simple sustainable ideas in the development of design (planning and creation), the lexical marriage of sustainability and design only began in the mid-1990s. In 1995, the Surrey Institute of Art & Design in the UK established the Centre for Sustainable Design which offered a master's course in sustainable design.[2] Researchers there proposed that sustainable design

---

[1]Vezzoli C, Kohtala C, Srinivasan A, et al. Product-service system design for sustainability[M]. Sheffield, UK: Greenleaf Publishing, 2017:2.

[2]The Surrey Institute of Art and Design, emerged in 1995 from West Surrey College of Art and Epsom School of Art and Design, later merged with Kent Institute of Art and Design in 2005 to become the University of Creative Arts.

X. Liu et al., *Design for Sustainability*, Sustainability and Interdisciplinarity, https://doi.org/10.1007/978-981-96-8657-5_2

involves analyzing and optimizing the system in which we create, use, and handle products, rather than merely engaging in local, short-term environmental design.[3]

In 1994, Emma Dewberry and Philip Goggin of the Open University in the UK suggested that, in contradistinction to the green design and ecological design, DfS must address more complex social, economic, and moral constraints beyond products. DfS thus involves shifting from products to systems, from hardware to software, from designing products to providing services.[4,5] At the beginning of the twenty-first century, Professor Ezio Manzini of the Polytechnico di Milano argued that DfS is distinct from general design, which simply produces material products. Instead DfS integrates products and services to construct sustainable solutions, a strategic design activity that aims to meet specific consumer needs, replace the consumption of physical products with results and benefits, reduce resource wastage, and mitigate environmental pollution, ultimately improving the quality of people's social lives.[6] By contrast, Tischner takes a broader perspective on sustainable design: he considers it an exploration of solutions that ensure benefits to society and surrounding communities (especially poor and underdeveloped ones), the natural environment, and the economic system (global, but especially local).[7]

DfS is a highly inclusive concept. While it inherits fundamental ideas and methods from green and ecological design, DfS practitioners do not simply seek to protect the environment. They advocate for a systematic innovation strategy that includes environmental and social benefits, users' needs, and corporate development. In this context, designers, who have traditionally played a service role, are increasingly assuming social responsibilities and obligations, thus expanding the territory and scope of traditional design.[8]

Design's development strongly depends on the social, economic, and cultural background in which it operates. Design as a means to an end and part of the service industry tends to be passive and behind the times. In what follows, we will discuss the awakening process of design in response to the urgency of sustainable development issues and briefly introduce the important concepts and evolution paths of integrating design and sustainability.

---

[3] Anne Chick proposed the concepts of system view and circular view of sustainable design in the Master's Degree course plan of the Sustainable Design Center in 1995.

[4] Dewberry E, Goggin P. EcoDesign and Beyond: steps towards sustainability [EB/OL]. [2021-10-12]. http://oro.open.ac.uk/29316/

[5] Dewberry E. Ecodesign strategies [EB/OL]. [2021-10-12]. https://ci.nii.ac.jp/naid/10012598857/en/

[6] Manzini E, Vezzoli C, Clark G. Product-service systems: using an existing concept as a new approach to sustainability[J]. Journal of design research, 2001, 1(2): 27–40.

[7] Charter M, Tischner U. Sustainable solutions: developing products and services for the future[M]. Sheffield: Greenleaf Publishing, 2001: 80, 204.

[8] Liu Xin. The concept, development and practice of sustainable design[J]. Creative Design, 2010 (2): 36–40.

2.1 The Awakening of Design

Historically, design has served as an important bridge that connects production, circulation, and consumption. After the Industrial Revolution, the exquisite craftsmanship of the handicraft era gradually disappeared along with luxury goods exclusively for the nobility. Replacing these were by mass-produced, rough but affordable industrial products. In reaction to the crude production quality and enslavement of workers this industrial production method entailed, John Ruskin and William Morris started the famous Arts and Crafts Movement, striving to make design more artistic and to revive handicraft traditions facing extinction. They also aimed to restore the joy in labor for workers through more creative work.[9] However, due to stubborn resistance to any form of mechanization, this idealistic design innovation movement did not thrive. Nevertheless, the philosophy and the exquisite works Ruskin and Morris left behind had a significant impact on subsequent design, including design for sustainability.

At the beginning of the twentieth century, members of the Bauhaus movement fully acknowledged and accepted industrialization, integrating technology and art into their work, and returning design to the functionalism that begins with human-centered principles, laying the foundation for modern design disciplines and industries. In the 1920s, the United States entered a period of rapid economic development, during which commercial design took center stage. Together with capitalism, commercial design created a wave of global consumerism. Because of their familiarity with human nature and their skill in manipulating aesthetic forms, commercial designers, under the premise of following the logic of capital, explored people's potential desires, launched innovative products, and maintained the operations and prosperity of consumer society by means of pervasive marketing.

We do not deny the positive contributions of commercial design to human. From the time Raymond Loewy appeared on the cover of *Time* magazine in 1949, design took the center stage and played an important role in social and economic development. However, for a long time, design for industry primarily played the role of a means, a tool for corporate profit. The responsibilities of designers seemed limited to creating dazzling, attractive products to promote sales and help companies make money, even as they turned a blind eye to environmental and social issues. From planned obsolescence[10] to formulaic product innovation competing for attention in the market, *design served as a direct manipulator which stimulated people's desires and shaped unsustainable views of happiness, consumption, and lifestyles. Consequently, design has received accusations of being a culprit in promoting consumerism, exacerbating resource consumption, and increasing waste output.*[11] In

[9] Charlotte Fiell, Peter Fiell. The Story of Design [M]. Translated by Wang Xiaomo, Wang Zhenshi. Nanjing: Jiangsu Phoenix Fine Arts Publishing House, 2018:142.

[10] For a detailed discussion on planned obsolescence, see Chap. 6 of this book.

[11] Liu Xin, Xia Nan. From End to Source: Waste Tracking and Product Service System Design [J]. Decoration, 2013(6):22–25.

1948, Edgar Kaufmann, Director of the Department of Industrial Design at the Museum of Modern Art in New York (MoMA), fiercely criticized the design industry for changing from "form follows function" to "style follows sales."[12] In his 2009 book, *Design is the Problem: The Future of Design Must be Sustainable*, Nathan Shedroff, a professor at the University of California, pointed out that while design should solve problems, it now contributes to creating them.[13] In light of this, the design community began to seriously reflect on its own purpose, which triggered the emergence of design for sustainability concepts.

When discussing the integration of design and sustainability in the present, we must first mention two heavyweights, namely Buckminster Fuller and Victor Papanek. Their thoughts and practices laid the foundation for today's design for sustainability.

In 1968, Buckminster Fuller, an American architectural designer, inventor, and environmentalist, published *Operating Manual for Spaceship Earth*. In this book, Fuller used the spaceship as an analogy to elaborate on the impact of human development on the environment, which was quite commendable at the time. Fuller realized that the way humans use resources could only meet the needs of a small portion of Earth's population, and most people were destined to experience poverty and suffering, which he considered extremely inefficient and foolish. His proposal was for a design revolution to eliminate flashy, impractical designs. Throughout his long life, Fuller continuously explored the impact of technology on human survival and development, forming a unique design theory called dymaxion, which entails a method of doing more with less energy. All his work adhered to this concept, including the famous Fullerene.[14] (Fig. 2.1). He once fantasized about covering all of Manhattan with it to create a self-sufficient ecosystem to resist possible future climate changes and man-made disasters, such as nuclear explosions.[15] In fact, we can still feel Fuller's design influence in many science fiction movies today.

Victor Papanek was an outstanding American designer and theorist and an ally and follower of Fuller. From the late 1960s to the early 1970s, the American economy was booming, and people working in the industrial and design sectors were full of confidence in the future. Despite this situation, Papanek proposed the concept of design ethics, uncompromisingly criticizing the rampant consumerism prevalent in the United States, as well as the sharp increase in urban energy consumption and

[12] Charlotte Fiell, Peter Fiell. The Story of Design [M]. Translated by Wang Xiaomo, Wang Zhenshi. Nanjing: Jiangsu Phoenix Fine Arts Publishing House, 2018:370.

[13] Shedroff. N. Design is the problem: the future of design must be sustainable [M]. Translated by Liu Xin, Qin Jingyan. Beijing: Tsinghua University Press, 2011: xxxv-xxxvi.

[14] The short-thread dome building Fuller designed for the US Pavilion at the 1967 Montreal World Expo, which is both lightweight and strong, was later named the Fullerene.

[15] If the earth's resources are completely depleted and the end of the world comes, what should we do? Fuller proposed a shocking Manhattan Dome plan in 1960. He envisioned using a Fullerene-style dome to cover the center of Manhattan, where the city could establish a complete, self-sufficient system. This dome would create suitable climate conditions, provide necessary ecological mechanisms, and have a complete waste disposal system. At the same time, the dome also has defensive capabilities and could protect Manhattan from solar storms or nuclear bombs.

Fig. 2.1 Buckminster Fuller and the US Pavilion at the 1967 Montreal World Expo (Image source: https://www.wired.com)

pollution. Papanek encouraged designers to pay attention to social issues, the third world, and the underprivileged and considered protection of the environment and natural resources part of designers' responsibilities. His classic work, *Design for the Real World*, was rejected by 12 publishers, with the first edition published in Sweden in 1970. The English version was published in the United States in 1971. Subsequently, the book was translated into numerous languages, revised and reprinted multiple times, and had far-reaching influence. In *Design for the Real World*, Papanek argues for self-restraint in industrial design, strongly criticizing commercial designs for consumers created purely for profit. Instead, he advocated designers take responsibility for social and ecological changes. His strong emphasis on the social and environmental consciousness of designers made subsequent generations of designers aware of the social and ethical values they should have, while his lifelong advocacy of the limited resources theory provided a theoretical basis for the later green design movement.[16] (Fig. 2.2)

Design as a never ceased to reflect and explore environmental and social issues. In the 1970s and 1980s, however, when Western manufacturing was developing rapidly, its social impact was extremely limited due to the dominance of commercial design. After the 1990s, with the continuous intensification of concern for environmental and social issues, the UN gradually raised awareness about sustainable

[16] Teng Xiaobo. Victor Papanek: Pioneer of Design Ethics [J]. Decoration, 2013(7): 60–61.

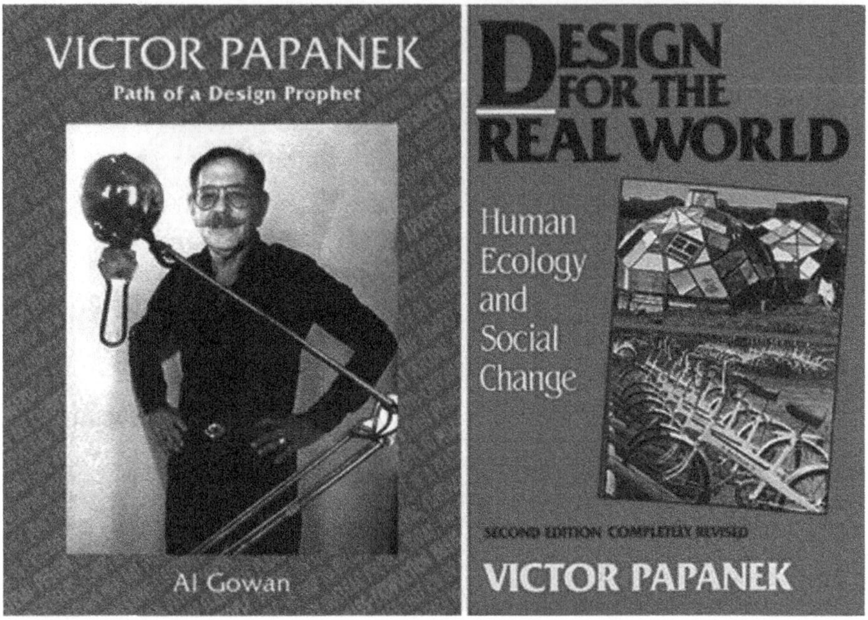

Fig. 2.2 Victor Papanek and "Design for the Real World." (Image source: Amazon & LinkedIn)

development through several declarations. The seeds Fuller and Papanek sowed slowly germinated, took root, and grew. The design community began to actively and proactively reflect on and adjust its role vis-à-vis society and established a systematic design theory system. For the design community as a whole, this has been a process of slow awakening. Green, ecological, and sustainable design were born and developed in this context, gaining widespread consensus and acquiring social attention. Now, they are mainstream to discourses in the design world.

This book seeks to reveal to readers the evolution and internal logic of this awakening, reflection, and action. Design as an independent discipline is still very young, influenced by various factors such as economy, society, culture, and technology. Coupled with the practicality of design and the basic characteristics of interdisciplinary hybridization, its evolutionary process has not followed a single linear trajectory, but instead has been the result of various developments. Moreover, as a result of emerging from different disciplinary fields, perspectives, and academic traditions, many sustainable design-related concepts have broad, approximate meanings that may overlap with each other. Although these problems are unavoidable, the internal logic of their development is still traceable. Based on many years of design research and accumulation of practical experience, in conjunction with the LeNS International Sustainable Design Learning Network project,[17] as well as the

[17] LeNS (the Learning Network on Sustainability) is an Asia Link project the European Union funded in 2007–2010 with the research theme of Sustainable Product Service System Design. It

research of scholars such as Carlo Vezzoli and Fabrizio Ceschin, the author briefly divides the evolution of sustainable design concepts into five stages: green design, ecological design, product-service system design, design for social fairness and harmony, and design for sustainable transformation. The following chapters will discuss these stages in depth.

2.2 Green Design

Starting in the 1970s, the number of severe environmental pollution incidents increased sharply, prompting the scientific community and the media to begin discussing the need to use materials and resources with low environmental impact[18] and to define standards for them. This marked the early stage of the green design movement, which aimed to reduce resource consumption and toxic emissions in the industrial sector using technical means. Understanding and mastering related material knowledge, technical standards, and rules became a necessary part of designers' knowledge base and design capabilities.

During this period, there was also public reflection and discussion on the need for natural materials. Natural materials as a term is vague and can create confusion. Compared to various synthetic materials, natural materials appear to be more environmentally friendly, often influencing consumer choices. However, many toxic and harmful substances exist in nature, meaning natural materials are not necessarily harmless. Furthermore, the process of making useful products from natural materials often requires complicated processing procedures, which can have significant environmental impacts (consider the emissions and hazards of the industrialized papermaking process). These reflections and controversies have played an important role in promoting the development of green design.

In the mid-to-late 1980s, discussion of green design deepened as public awareness of ecological disasters and environmental issues continued to rise, and the

---

involved seven design institutions in Asia and Europe and aimed to promote exchange and cooperation among higher design education institutions in design for sustainability research, teaching, and practice through the establishment of an open network learning platform. The LeNS project led to the formation of several regional sustainable design alliances which jointly applied for several international sustainable design research projects, such as LeNSes (the Learning Network for Sustainable Energy Systems), with the research theme of Design Research on Sustainable Energy (2013–2016); and LeNSin (the International Learning Network of networks on Sustainability), focusing on Sustainable Product-Service System Design and Distributed Economy (2016–2020). See http://www.lens-international.org

[18] Environmental impact refers to the effects and changes to the environment human activities (economic activities, political activities, and social activities) cause, as well as the effects on human society and the economy resulting from these changes. Environmental impact likewise includes the environment's reaction to humans. See: Wang Zhe, Wu Ben. Environmental Impact Assessment[M]. Tianjin: Nankai University Press, 2014:2–4.

Green Party came to prominence in Europe.[19] A large number of green-themed images appeared in the media and advertisements, bringing new symbolic meanings to design and leading to the emergence of buzzwords such as green design, green products, and green packaging. *During this period, green design became synonymous with design to mitigate environmental impact.* In 1986, the British Design Council held a design exhibition called Green Designer, which for the first time explained to designers and the public what green design is in terms of energy use, durability, recyclability, and market acceptance. The exhibition proposed the slogan, "Win in Design, Profit from Design." Shortly thereafter, the term green consumer was first used to describe a new trend in consumption, indicating that green design is not anti-industrialization but can help industries progress. In 1991, the Design Council published the book Green Design, marking the beginning of comprehensive and systematic research on sustainability in design.

As a concept, green design encompasses industrial product design, clothing design, information and visual communication design, environmental art design, architecture, and planning. Regardless of the field, green design shares some basic principles. In 1994, scholar Paul Burall proposed that green design should follow the 4Rs: reduce the use of materials and energy; reuse old products or parts to reduce waste generation; recycle waste materials; and regenerate by turning waste into valuable products or raw materials.[20] In 1994, Mackenzie proposed six operational principles for green design: (1) easy disassembly and assembly to increase the convenience of maintenance; (2) use of low-pollution and easily recyclable materials; (3) complete environmental protection markings; (4) clear recycling classification; (5) consideration of recycling value; and (6) effective waste disposal.[21] In 2002, Duriez suggested that green design should reduce natural resource consumption, select the most suitable materials for the ecological environment, use the least polluting production processes, and increase product usability and lifespan. Since then, strategies such as design for disposal, design for disassembly, design for durability, and the use of biodegradable alternative materials have become key elements of green design.

Green design was the first to incorporate environmental issues as core elements of design thinking, marking profound reflection on the role and social responsibility of design and significantly enhancing its social value. Green design represents not a change in design style, but rather an evolution in design attitude, concept, and

---

[19] The Green Party is an international political party advocating for green politics. Green politics has three basic goals: pacifism, social justice (especially emphasizing the rights of local residents), and environmental protection. Supporters of the Green Party believe that the realization of green politics can make the world healthier and they often take action to achieve their ideals, opposing any economic development strategies that cause ecological damage. Source: Wikipedia.

[20] Burall P. Green-ness is good for you[J]. *Journal of Product Innovation Management*, 1995, 12(4): 354.

[21] Mackenzie D, et al. Green design: design for the environment[M]. London: Laurence King, 1991:154.

corresponding strategies, prompting deep reflection on the relationship between economic development and the environment.

However, *early green design focused on reducing the environmental impact of materials and components, that is, post-process intervention. After recognition of a problem and the harm it does came the implementation of remedial measures to mitigate impact. Essentially, this approach only reduced the intensity of harm to a certain extent and prolonged the ongoing cycle of harmful effects.* Advances in understanding and technology have however continuously refined and improved the concept of green design. Over the next few years, its boundaries expanded, leading to broad discussions and applications in various product fields, with ecological design emerging as a more appropriate term.

2.3 Ecological Design

In the mid-1990s, researchers steadily shifted their focus from localized questions of technical-level materials and energy concerns to the level of complete products. The analysis and design of a product's lifecycle became central to their work. They called this new method, rooted in understanding natural systems and ecology, ecological design or eco-design, for short. *Ecological design encompasses almost all the same strategies and principles as green design, and the differences between these concepts are not absolute.* Indeed, the specific terminology matters less than the core idea it expresses. Ecological design practitioners consider and optimize the environmental impact of a product's lifecycle from a macro perspective, offering an inclusive approach that has philosophical implications. In contrast, green design has a clearer, more specific entry point and is more operational.

From an etymological perspective, ecology, which German doctor and zoologist Ernst Haeckel coined in 1866, is a branch of science that analyzes the relationships between organisms and their environment. In the 1960s, activists fighting environmental pollution adopted the term.[22] In 1982, the German scientist Evelyn Möller proposed the term ecological functionalism and created an ecological checklist for German industrial designers and manufacturers, which formed the basis for the German Industrial Design Association (Verband Deutscher Industrie-Designer, VDID) to promote ecology and design.[23] In 1989, the UK established the Eco-Design Association and founded the journal *Ecodesign*. This specialized term reflects the European design community's understanding of the relationship between design and ecology from the perspective of deep ecology.[24]

---

[22] Quoted from Word Origins Dictionary (https://www.quword.com/etym/s/ecology).

[23] Madge P. Ecological design: a new critique[J]. *Design issues*, 1997, 13(2): 44–54.

[24] For the meanings of deep ecology and shallow ecology, see Chap. 1.

"The design of materials and products, projects and systems, environment, community, should be friendly to the ecological environment and all life forms of this planet."[25] In the 1990s, the concept of ecological design quickly spread from Europe to the rest of the world, formulating a new value system guiding design development.

In ecological design, the environment receives equal status with traditional design values, such as functionality, commercial benefits, aesthetics, and ergonomics. Practitioners of ecological design have developed a series of complete principles, guidelines, and toolkits. A notable feature of the methods they use includes viewing the environmental impact of products from a full lifecycle perspective, calculating total energy consumption and environmental impact from raw material acquisition and transportation through production, sales, use, and disposal. Using quantitative analysis, ecological designers identify the stages and activities posing significant problems, leading to targeted design interventions. For more on this, see Chap. 3, Lifecycle Design.

Compared to green design, the full-process analysis inherent to ecological design is more systematic and comprehensive. However, achieving a truly objective and accurate product lifecycle assessment is extremely difficult, requiring extensive data collection and years of tracking. This raises questions about whether early assessments of products' lifecycles involved scientific methods or theoretical assumptions. Recent advances in computer algorithms and databases have, however, improved the credibility of lifecycle assessments.

Despite these challenges, the influence of ecological design is profound, informing a wide range of theories and methods such as industrial ecology, cradle-to-cradle (C2C) circular design, permaculture, and biomimicry design.

Eco-design is a design approach that addresses the entire product lifecycle, not merely focusing on the final result but comprehensively considering environmental issues at every stage, aspect, and phase of the product's development. It is essentially an "intervention within the process." However, in general, both green design and eco-design concentrate on the design of materialized products, with the core objective of reducing environmental impact. They lack a comprehensive consideration of social issues and business models, as well as the ability to propose systematic and sustainable solutions.

The concept of eco-design has also faced broader challenges.

Emma Dewberry and Phillip Goggin argue that sustainable design must address complex social, environmental, development, and ethical issues. They assert that the current development model is extremely unfair to developing countries, since 20% of the population in developed countries consumes 80% of the world's resources. Consequently, ecological design, which focuses on the relationship between the environment and economic development, has significant limitations and cannot achieve global sustainable development.[26] Gui Bonsiepe further contends that eco-

---

[25] Quoted from the Ecological Design Concept leaflet published by the Ecological Design Association (EDA), UK, in 1990.

[26] Dewberry E, Goggin P. EcoDesign and beyond: steps towards sustainability [EB/OL]. [2021-10-12]. http:/oro.open.ac.uk/29316/.

logical design remains the luxury of wealthy countries, with people in the Global South absorbing the costs of environmental standards in the Global North.[27]

2.4 Product-Service System Design

In the late 1990s, a series of research reports from the United Nations, the European Union, and other institutions deepened our understanding of sustainability, revealing that technical measures targeted at the material level were insufficient. Achieving sustainable development requires changing business models, consumption patterns, and implementing system-level changes. Early green design and ecological design faced significant implementation challenges. In traditional design models focused on selling products, companies aimed to reduce material and energy consumption during production but neglected the energy consumption of products during use and their subsequent disposal. This approach often resulted in resource waste and environmental impact from recycling, transportation, and processing. Durability became a mere marketing slogan, as manufacturers preferred to sell products with short lifecycles to boost sales and profits. Some companies used greenwashing,[28] taking advantage of the lack of unified standards and consensus over what constitutes green and ecological, as a marketing gimmick. Traditional profit-driven enterprises lacked the genuine motivation to invest in environmentally friendly product design and manufacturing.

Even when design strategies yielded environmental benefits, increasing product consumption often offset these gains. For example, the European Union reported that the environmental benefits of improving car efficiency by 10% from 1990–2004 were negated by a 30% increase in car sales and mileage. Thus, relying solely on product-level lifecycle design and ecological design proved insufficient. This realization led many designers and scholars to advocate for fundamental changes to traditional business models, emphasizing innovation beyond the development of different kinds of material products. Consequently, focus shifted to the design of non-materialized product-service systems (PSS).

From an enterprise perspective, PSS design represents an innovative strategic shift from merely designing and selling material products to providing a comprehensive product and service system to better meet people's needs. This approach changes the business focus from profiting from the sale of numerous products to providing a sense of fulfillment by meeting customer requirements. Traditional manufacturing industries, facing market shrinkage and intensified competition, began to view services as a new avenue for profit growth. They recognized that combining products and services could yield higher profits than selling each

[27] Madge P. Design, ecology, technology: a historiographical review[J]. Journal of design history, 1993, 6(3): 149–166.

[28] Greenwashing generally refers to deceptive, profit-driven environmental propaganda and actions.

separately. This shift in perspective led to the development of the product-service system concept, blending and evolving product and service integration.

From a consumer perspective, the actual need is for a satisfactory outcome rather than specific functional products. For instance, people need clean clothes rather than washing machines and functional urban transportation requires convenient travel options besides private automobiles. Products are merely tools for achieving these goals. Therefore, designers must transition from product design thinking to system and service design thinking. *Properly implemented, PSS innovation could effectively reduce resource consumption, lower enterprise costs, and provide new profit growth opportunities, while addressing consumer needs. More importantly, PSS can promote a non-material consumption concept, establishing new lifestyles and consumption patterns, and reduce the environmental impact of human activities at the source.* The sharing economy's principle of use rather than own has thus become a crucial strategy and principle of PSS design.

The design of sustainable product-service systems (SPSS) aims to integrate various design-related factors into a complex business environment, focusing on the interaction patterns of different stakeholders and creating comprehensive solutions for new product-service business and consumption models. As Paul Hawken states in *The Ecology of Commerce*, enterprises need to unify economic, biological, and human systems into a whole, achieving a cycle of coexistence between businesses, consumers, and the ecological environment, thereby paving the way for sustainable business development. The painful process of economic development involves shifting from over-reliance on the production and consumption of material products. This transformation is essential for the future Chinese and global economy to achieve sustainable development. For more on product-service system design, see Chap. 9.

2.5 Design for Social Equity and Harmony

After 2000, the concept of design for sustainability expanded to address social issues, with a particular focus on the needs of developing countries and people at the base of the pyramid. Notably, the design concept of social concern predates the concept of sustainable development. As early as the 1970s, scholars recognized that designers should bear greater social responsibilities than they did. For instance, Thomas Maldonado's Design Hope movement and Victor Papanek's assertion that "design can and must become a means for young people to participate in and change society" were pivotal ideas, albeit largely excluded from the mainstream discourse of design.

A significant concept, Design for the Base of the Pyramid (DfBP), has much in common with the early concept of social design. Both are reflections on design stimulating consumerism and only serving the commercial design for the elite. DfBP has evolved through two stages: the first focused on addressing the survival needs of those at the bottom of the pyramid and using participatory methods to empower them; the second stage treated these individuals as business partners, enabling their empowerment through business co-creation. Over time, DfBP has broadened its scope to include business models and complex social ethics.[29]

The democratization of design is another crucial topic in social design, prompting reflection on who truly benefits from design services and even the question of who qualifies as a designer. This deepens consideration of the relationship between design elites and the general public.

In the twenty-first century, scholars like Margolin, Manzini, Tischner, and Vezzoli argued that design for social equity and cohesion is vital to sustainable development, which marked an expansion of design for sustainability. Subsequently Design for Social Innovation and Sustainability became a research hotspot. As part of this, Professor Manzini promotes the Design for Social Innovation and Sustainability Network (DESIS) as a global platform for research and practice. He argues that social innovation stems from novel solutions that ordinary people devise to address daily problems, often relying on social capital, which includes connections, trust, and cooperation. Encouraging spontaneous public participation and cooperation can lead to new social structures and systemic innovation. In this framework, professional designers shift from acting as leaders to being collaborators, thus redefining the discipline of design. Manzini envisions a future of design as green, social, and networked. For more details, see Chap. 10.

Achieving sustainable social and economic development fundamentally requires transforming people's values and consumption views. Implementing sustainable consumption patterns is crucial to the success of social dimension design. Since design bridges production and consumption, it can either stimulate consumption impulses or advocate for sustainable consumption. Intervening in consumer behavior, especially during and after consumption, is essential. Design for sustainability must link information, individual behavior, and impacts on the environment and society. This need has given birth to design for sustainable behavior.[30] Dr. Debra Lilley from Loughborough University argues that design-led methods should influence user behavior to reduce negative social impacts during product use. She explores that design can shape user behavior for sustainability, aligning with the emerging field of behavioral economics.

In summary, social sustainability depends upon how individuals, communities, and societies interact. It requires choosing development models that take into consideration local and global environmental capacities. Design for social sustainability represents the frontier of current design research, expanding and improving on sustainability as a concept. It addresses areas that extend well beyond environmental and economic sustainability, such as social fairness, welfare, democratic participation, cultural and species diversity, and sustainable consumption. A systemic concept, it deepens and perfects design for sustainability, focusing on social

[29] Ceschin, F., Gaziulusoy, I. *Design for sustainability: a multi-level framework from products to socio-technical systems* [M]. London: Routledge, 2019:88.

[30] Lilley, D. "Design for sustainable behaviour: strategies and perceptions[J]." *Design Studies*, 2009, 30(6): 704–720.

harmony in a globalized world and the spiritual and emotional well-being of the public.

2.6 Design for Sustainable Transformation

In recent years, the social, environmental, and economic problems humanity faces have become increasingly complex and interrelated. Consequently, design, which is fundamentally about discovering and solving problems, has begun to engage with the deep levels of changes to socio-technical systems which innovations in technology, society, organizations, and mechanisms trigger.[31] To address complex problems, such as climate change, environmental pollution, crime, and poverty, scholars like Terry Irwin, Cameron Tonkinwise, and Gideon Kossoff from Carnegie Mellon University have proposed the concept of transition design. They aim to promote the transition of society toward a future more sustainable and ideal than the present. *This indicates that design for sustainability is no longer limited to innovations in materials, products, and business models, but now concerns (although currently to a limited extent) engagements in urban renewal, system reconstruction, social network reshaping, and broad social transformation.* Terry Irwin posits that *transition design is an emerging field within design for sustainability, following service design and social innovation design.* Transition design includes ideas from product-service system design, which aims to transform production and consumption patterns through business model innovation, and social innovation design, which seeks to encourage social change without relying on technological change as a prerequisite.

The Transition Town Movement, that British environmentalist Rob Hopkins initiated, inspired the concept of transition design.[32] Additionally, beginning in the late twentieth century, scholars focused on theories and methods of system innovation from a multi-level perspective, laying the foundation for the emergence of transition design. Brezet (1997) was one of the first scholars to discuss this topic within the context of design, defining system innovations as changes in infrastructure and organization with the goal of meeting the requirements of new products and servic-

---

[31] E.L. Trist and his colleagues at the Tavistock Institute (UK) proposed the concept of socio-technical systems. They propose that an organization or enterprise is both a social system and a technical system, with the technical system having a significant impact on the social system. For instance, the technical system significantly influences individual attitudes and group behaviors. A major task for managers is to ensure that these two systems coordinate with each other. Broadly speaking, this theory also elucidates relationships between people's ideas, behaviors, and technological development in society.

[32] Transition Town Movement is a grassroots movement that originated in Europe, often carried out on a town-by-town basis, for systematic innovation. The small town of Totnes in southeastern England is considered the first transition town. It issued local currency pegged to the pound, developed local planting and food networks, encouraged local industries and consumption, and supplemented public services through community collaboration such as cooperatives. Currently, this informal network has over 300 members worldwide.

es.[33] This understanding surpasses traditional design boundaries, involving mechanism innovation and strategic design

In 2008, the first international conference on design and sustainable transformation, Changing the Change, was held in Turin, Italy. The conference participants emphasized that ... We must take the term sustainability for what it effectively means—a radical change in ways of being and doing... Sustainability has to be the meta-objective of all design research.[34]

In 2012, Carnegie Mellon University integrated the concept of transition design into its undergraduate, graduate, and doctoral programs, proposing a theoretical framework that explores design's role in transitioning to a sustainable society. This framework includes four interrelated aspects: (1) visions of transition: establishing a clear direction; (2) theories of change: explaining complex systems and their inherent changes; (3) mindset and posture: adopting an open, collaborative, introspective stance; and (4) new ways of designing: developing new design methods informed by the previous three aspects.

Specifically, the vision of transformation requires participation from all stakeholders to reimagine new ways of life and establish a future-oriented vision. Based on a dynamic understanding of complex system changes, this vision promotes the creation of a new knowledge system, fostering a deep comprehension of social and natural systems' dynamics. This interdisciplinary approach reshapes designers' knowledge structures, thinking modes, and positions, guiding the development of new design methods to achieve the envisioned transformations. This process is bidirectional, with the evolving vision encouraging continuous improvement and iteration of design methods, further refining designers' knowledge and approaches. Transition designers focus on three broad areas: establishing strong narratives and visions for the future, expanding and connecting local communities and organizations, and designing innovative localized solutions using interdisciplinary teams.

In addition to transition design, sustainable thinking has integrated new discourse systems, gaining fresh connotations and meanings. This includes speculative design,[35] design activism, design for resilience, design for biodiversity, regenerative design, and explorations integrating feminist theory, animal studies, post-humanist ethnography, and political ecology into sustainability. Speculative design encourages designers to think about future possibilities by integrating design with new theories, technologies, or experiments. It is not only about envisioning possible futures but also urging people to clearly recognize the current state and challenges of technological development. *It emphasizes discovering problems, functional fictions, and exploring how the world develops—fostering new perspectives and promoting debate, thinking, and decision-making around controversial issues.*

---

[33] Ceschin F, Gaziulusoy I. *Design for sustainability: a multi-level framework from products to socio-technical system*[M]. London: Routledge, 2019:125.

[34] Ciuccarelli P, Ricci D, Valsecchi F. Changing the change: design, visions, proposals and tools[C]. Turin: Allemandi Conference Press, 2008: 5, 22.

[35] Anthony Dunne and Fiona Raby, who previously proposed critical design, which also has significant influence, proposed the concept of speculative design.

In an uncertain and changing world, designers can explore possible futures through imagination and speculation, as well as concrete actions. Design activism encourages designers to engage with social movements, mutual aid, and marginalized communities, providing necessary support through a spirit of inclusion. The value of design activism lies in promoting social change, raising awareness of values and beliefs (such as climate change and sustainable development), and challenging the constraints of mass production and consumerism. Alastair Fuad-Luke describes design activism as critical imagination, "Design can rupture the present with counter-narratives…Design is critical imagining. Design generates considered possibilities for a new, 'beautiful strangeness' (with new values embedded or implicit). Design can readjust our notion of beauty to embrace a multitude of truths—economic, political, social, ecological, ethical, technical, symbolic, institutional, philosophical, and cultural."[36] As adherents to a relatively radical concept, design activists often refuse to cooperate with corporations and other commercial institutions, aiming to counteract capitalist development through critical design planning. Design activism is not a political act, but an intervention in life from the designer's perspective, enhancing the reflective power and influence of actions to choose a sustainable future.

2.7 The Inherent Logic of Evolution

Examining the development of design for sustainability, we observe a progression from green design and ecological design, which focus on environmental issues, to product-service system design involving innovations in business and consumption models, to social innovation design, which addresses social equity and behavioral changes, and finally to transition design aimed at developing a sustainable future. This progression reflects an increasing complexity in problem-solving, a shift from end-of-pipe treatment to source intervention. This trajectory mirrors human approaches to environmental challenges. Since the mid-twentieth century, as the ecological crisis has become increasingly evident, our responses evolved from pollution control (treatment of industrial pollution) to process intervention (employing clean technologies to reduce emissions), to the design of new products or service systems (systemic change), and ultimately to fostering ecological consciousness and societal behavior changes (sustainable lifestyles and consumption). *Over time, the role of design in sustainable development has become increasingly pronounced and fundamentally transformed—from being a form supplier to becoming a coordinator of various interests, a proactive agent, and providing visions for a sustainable future.*

Based on Fabrizio Ceschin's research, Fig. 2.3 delineates the evolution of sustainable design concepts into five stages, positioning the themes covered in this book (with the exception of transition design). It shows the evolution from a

[36] Fuad-Luke, Alastair. *Design activism: Beautiful strangeness for a sustainable world.* Routledge, 2013: xxi

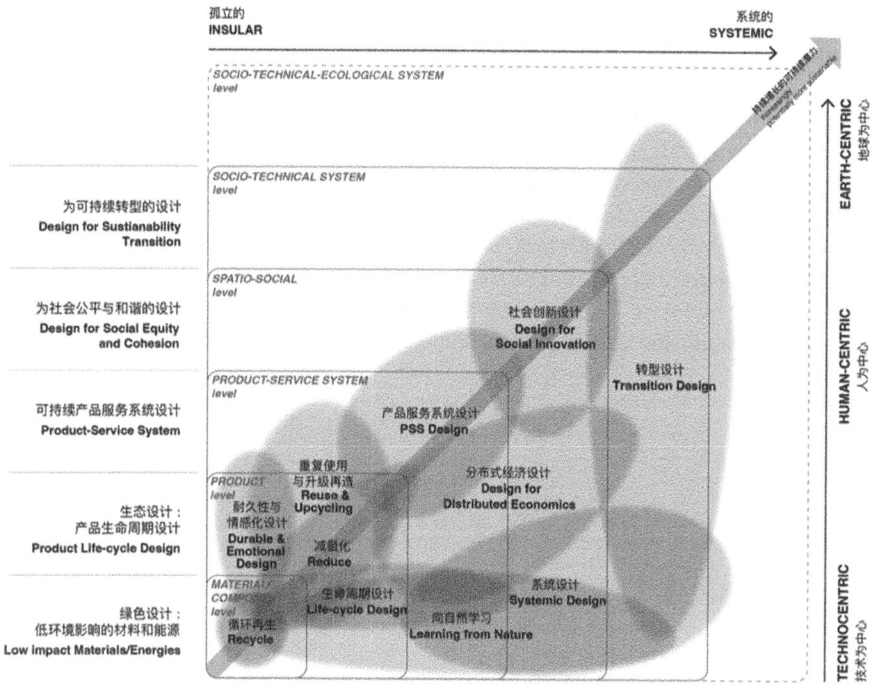

Fig. 2.3 Evolution trend and research themes of sustainable design. (The framework structure is adapted from Fabrizio Ceschin's research at Brunel University, UK)

relatively isolated, technology-centered perspectives to a more systemic, human-centered, and ultimately earth-centered perspectives. As discussed in Chap. 1, while being earth-centered (or eco-centered) reflects upon and serves as a correction of obsessive anthropocentrism, instead considering humanity's long-term welfare, it fundamentally remains a human-centered view. The figure clearly illustrates that integrating systematic and earth-centered perspectives has increased potential for sustainability, serving a focus of future sustainable design research.

Nevertheless, isolated, technology-centered perspectives still hold significant value, as they are the foundation for designing man-made objects. Basic strategies such as circular regeneration, reuse, upcycling, durability, and emotional design remain crucial topics in sustainable design today. Technological innovations continue to infuse new energy and inspiration into design. Biomimetic design, which uses advanced technologies to study biomaterials, structures, principles, and ecosystems, exemplifies this by opening new horizons for sustainable design innovation. Moreover, advancements in the capacities of big data and artificial intelligence provide new opportunities for designing for sustainability, especially when it comes to transition design, where the transformation of socio-technical systems relies heavily on profound understandings and ingenious uses of breakthrough technologies. Even so, transition design, which includes new knowledge, relationships, experiments, and challenges, lacks mature theoretical frameworks and successful case studies. Consequently, this book does not include a chapter dedicated to transition design, which will need further in-depth research and supplementation in the future.

The analysis in this chapter enhances our understanding of the spirit and content of design for sustainability. It is important to note that t*he divisions into different stages reflect a summary overview of the development of design for sustainability. These divisions do not imply that subsequent stages replace previous one. Instead, a continuous process of reflection, enrichment, and improvement occurs.* As shown in Fig. 2.3, many theories and strategies in sustainable design are interconnected and mutually inclusive. In the following chapters, we will embark on an exploration of various sustainable design themes named above

Chapter 3
Lifecycle Design

Abstract Lifecycle design of a product encompasses a range of strategies that treat the product from a temporal perspective and intervene at a systemic level. This approach aims to minimize a product's overall environmental impact throughout its entire lifecycle, from raw material extraction to waste disposal. The fundamental principle of lifecycle design is to move beyond end-of-pipe solutions by addressing manufacturing at the source and focusing on system, holistic, and process-oriented design.

Keywords Product lifecycle · Cradle-to-cradle · Functional unit

Similar to natural life forms, which undergo various stages such as birth, aging, sickness, and death, artifacts also have lifecycles. *From a sustainability perspective, lifecycle design applies temporal logic and systemic intervention, meaning that design must intervene in the entire lifecycle of a product, from the extraction of raw materials to the disposal of waste, aiming to reduce the overall environmental impact by minimizing resource and energy consumption, as well as waste and pollution emissions.*

Forward-thinking designers began emphasizing the need for environmental responsibility in the mid-twentieth century. Notably, in the 1980s, Dieter Rams highlighted harmony with the environment and innovation as key principles of good design.[1] Today, major international design competitions recognize environmental considerations as fundamental in major international design competitions, with numerous ecological, green, and low-carbon design strategies and guidelines emerging. However, localized or single design improvements, especially end-of-pipe solutions,[2] are limited in their capacity to reduce overall environmental impact. Therefore, adopting a holistic lifecycle system design perspective is crucial.

[1] The ten principles of good design Dieter Rams proposes are: Good design is innovative, it makes the product useful, it is beautiful, it makes the product easy to understand, it is unobtrusive, it is honest, it is long-lasting, it is consistent in detail, it is environmentally friendly, and it is as little design as possible.

[2] End-of-pipe solutions are a series of pollution control measures implemented at the end of the industrial production process, aiming to reduce impact on the environment. It is also used metaphorically for non-fundamental (source) remedial measures.

X. Liu et al., *Design for Sustainability*, Sustainability and Interdisciplinarity,
https://doi.org/10.1007/978-981-96-8657-5_3

The significance of lifecycle design lies in its proactive incorporation of environmental awareness from the very outset of a project. This approach helps prevent and mitigate problems rather than addressing damage at the final stage, which often incurs great costs in terms of time, health, and money.[3] This makes lifecycle design both environmentally and economically advantageous.

In essence, lifecycle design is a collection of strategies, the philosophy of which we will elaborate upon in subsequent chapters in the following sequence: recycling aims to maximize a material's lifecycle; reuse and upcycling seek to extend a product's lifecycle; durability design and emotional design prevent premature disposal of a product; and reduction fundamentally decreases consumption of resources and waste generation, closely aligning with a source-focused environmental design strategy. Additionally, product-service system design, biomimicry, and systemic design are all integral components of lifecycle design. This approach ensures that design incorporates sustainability at every stage, fostering innovative solutions that contribute to a sustainable future.

3.1 Back-to-the-Source Design

The linear design development model of resources to products to waste dates from the 1950s, when the concept of product lifecycle first appeared in marketing campaigns that divided it into four stages: development, growth, maturity, and decline. Marketers used this framework to describe the market performance trajectory of a product. The purpose of laying out these stages was to encourage companies to formulate corresponding competitive strategies tailored to each product stage, thereby maximizing profitability.[4]

In recent decades, there has been growing awareness of the drawbacks of this linear development model. This awareness has led to a realization that learning from and coexisting with nature is the correct path to sustainable development. As a result, the concept of product lifecycle has changed to one that originates from an understanding of natural laws: the life course of a biological entity is a process of continuous energy, material, and information exchange with the external biological and non-biological environment. All biological entities experience a complete lifecycle from birth, growth, decline to death. Similarly, any man-made object also undergoes a life course. This concept of the product lifecycle entered marketing in the mid-20th century and thereafter engineering.[5]

---

[3] Vezzoli C. Design for environmental sustainability: Life cycle design of products [M]. 2nd ed. Berlin: Springer, 2018:37.

[4] Levitt T. Exploit the product life cycle [EB/OL]. [2020-03-15]. https://hbr.org/1965/11/exploit-the-product-life-cycle

[5] Cao H, Folan P. Product life cycle: The evolution of a paradigm and literature review from 1950—2009 [J]. Production Planning and Control, 2012, 23(8): 641–662.

In the field of engineering design, the concept of the product lifecycle includes the entire process from raw material acquisition, manufacturing, use to disposal. Initially, this concept aimed to help companies control the costs associated with acquiring materials. In 1962, designer Morris Asimow proposed a circular product lifecycle model, which includes production, distribution, consumption, and recycling/disposal. This model emphasizes the recycling of waste back into production to improve resource utilization.[6] The inspiration for this concept comes from the symbiotic relationships in ecosystems, where the waste of one organism often serves as food or nutrients for another. Although the term industrial symbiosis appeared as early as the 1940s, Asimow's contribution was to build an ideal product lifecycle model based on this idea and propose a rudimentary design strategy (Fig. 3.1).[7]

Likewise in the 1960s, lifecycle assessment[8] (LCA) began to emerge as a topic of scientific research. LCA employs quantitative methods to analyze the material and energy flows of a product from raw material acquisition to final disposal, thus scientifically presenting resource consumption and environmental impact. LCA played a significant role in updating knowledge systems in engineering and spurred a qualitative change in design methods aimed at reducing the environmental impact of products. The Coca-Cola Company was the first industry to use the LCA method for product management.[9]

In the 1980s and 1990s, with the maturation of LCA methods, green design gradually evolved into eco-design, which comprehensively considers resource consumption and environmental impact throughout a product's entire lifecycle.[10] This

[6] Asimow M. Introduction to design [m]. Englewood cliffs, NJ: Prentice-hall, 1962:8.

[7] Early literature on the concept of industrial symbiosis: Renner g t. geography of industrial localization[j]. economic geography, 1947,23(3): 167–189. cited from: Chertow M, Ashton W, Kuppalli R. The industrial symbiosis research symposium at Yale: Advancing the study of industry and environment[C/OL].(2004-11-01) [2021-09-16]. https://elischolar.library.yale.edu/fes-pubs/23

[8] The national standard GB/T 24040—2008, which benchmarks the international standard ISO 14040:2006, defines lifecycle assessment as "the compilation and evaluation of the inputs, outputs and potential environmental impacts of a product system throughout its lifecycle."

[9] In the face of energy crisis and environmental pressure, the Coca-Cola Company in the United States commissioned the Midwest Research Institute in 1969 to conduct a comparative study on glass bottles and plastic bottles, aiming to optimize production processes and adjust business strategies. The study quantitatively analyzed the inputs (material and energy consumption) and outputs (waste emissions) of both during their entire lifecycle (from raw material acquisition to waste disposal). The results showed that although glass materials are widely sourced and the processing technology is relatively simple and mature, the consumption of materials and energy in a glass bottle's lifecycle leads to higher costs, especially during transportation; in comparison, plastic materials are lightweight, durable, and easier to recycle. As a result, Coca-Cola Company concluded that plastic is more environmentally friendly. Although this conclusion is still highly controversial today, this study pioneered the lifecycle assessment: People began to re-examine and understand products from the system perspective of the lifecycle.

[10] Ceschin f, Gaziulusoy I. Evolution of design for sustainability: From product design to design for system innovations and transitions [J]. Design Studies, 2016, 47:118–163.

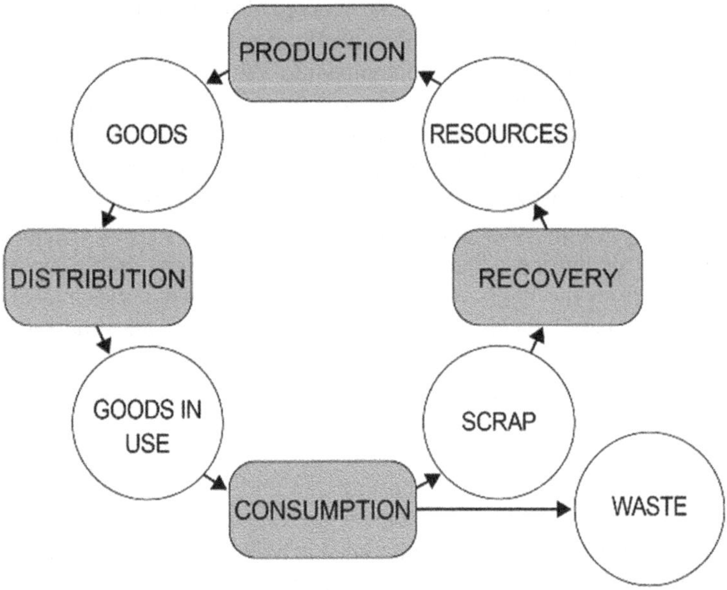

Fig. 3.1 Morris Asimow's production–consumption cycle (Asimow M. Introduction to Design [M]. Englewood Cliffs, NJ: Prentice-Hall, 1962:8)

transition from green to ecological design reflects the integration of engineering and ecology disciplines.

In 1989, Robert Frosch and his colleagues proposed mimicking the functions of natural ecosystems to establish an industrial ecosystem that "makes full use of raw materials and energy, produces as little waste as possible, and makes waste a raw material for another industrial process."[11] This laid the theoretical foundation for the interdisciplinary field of industrial ecology. Eco-design and LCA are integral to industrial ecology as key design and evaluation methods, respectively.

The 1990s saw the emergence of various environmental design methods in engineering, all based on the lifecycle framework. These methods aimed to reduce the environmental impact of products through optimization of functions, materials, and structures and incorporated LCA methods. Through mutual learning and influencing among designers, these approaches—such as design for the environment, lifecycle design, environmentally conscious design, and eco-design—began to converge.

At the beginning of the twenty-first century, nature increasingly inspired research and innovation. In 2012, McDonough and Braungart published the influential book, *Cradle to Cradle: Remaking the Way We Make Things*, in which they rethought creation activities in the artificial world. The cradle-to-cradle concept advocates for

[11] Frosch R A, Gallopoulos N E. Strategies for Manufacturing [J]. Scientific American, 1989, 261(3): 144–152.

regeneration, rather than consumption, in industry. It is based on lifecycle but emphasizes the upgrading of materials. Biological materials (e.g., bamboo) are designed to enter an open-loop ecological system for recycling after their service life ends, while technical materials (e.g., plastics) circulate within a closed-loop industrial system, avoiding waste generation and nourishing the system. This enables sustainable production, consumption, and economic growth.[12,13]

A key concept in cradle-to-cradle design is eco-effectiveness, which differs from the previous pursuit of eco-efficiency.[14] Eco-efficiency focuses on minimizing material flow within a product's lifecycle, seeking to optimize input–output ratios through technical means. However, eco-efficiency maintains the linear path from cradle to grave, entailing a decline in material quality. In contrast, eco-effectiveness maintains or enhances the quality of materials through a cyclical cradle-to-cradle process, akin to what occurs in a biological metabolism. *Eco-efficiency is about doing things in the right way, while eco-effectiveness is about doing the right things.* These strategies complement each other, leveraging their respective advantages.[15] The cradle-to-cradle concept is a foundational element of the widely discussed circular economy[16] and the blue economy[17] models (Fig. 3.2).

---

[12] Mcdonald, Braungart. Exploration of circular economy design from cradle to cradle [m]. China twenty-first century agenda management center, China-US center for sustainable development, trans. Shanghai: Tongji University Press, 2005.

William Mcdonough, Michael Braungart, cradle to cradle: Remaking the way we make things

[13] Ceschin F, Gaziulusoy I. Evolution of design for sustainability: From product design to design for system innovations and transitions [j]. Design studies, 2016, 47:118–163.

[14] The world business council for sustainable development officially proposed the concept of ecological efficiency in 1992. The organization for economic co-operation and development defines ecological efficiency as "the efficiency of ecological resources in meeting human needs," further stating it as "a ratio of output (the value of products and services produced by a company, sector, or the entire economy) to input (the total environmental pressure produced by a company, sector, or economy)." Braungart and others generally understand it as "obtaining more product or service value with less waste, less resource use, or less toxicity." (oecd.eco-efficiency[r/ol]2008 [2021-09-16]https://read.oecd-ilibrary.org/environment/eco-efficiency_9789264040304-en#page7)

[15] Braungart M, Mcdonough W, Bollinger A. Cradle-to-cradle design: Creating healthy emissions-a strategy for eco-effective product and system design [j]. Journal of cleaner production, 2007, 15(13–14): 1337–1348.

[16] The Ellen MacArthur Foundation report "toward the circular economy: accelerating the scale-up across global supply chains" calls the circular economy "an industrial system that is restorative or regenerative by intention and design. Its aim is to eliminate waste through the superior design of materials, products, systems, and business models." Ellen MacArthur foundation, toward the circular economy vol. 3: accelerating the scale-up across global supply chains[r/ol] 2014[2021-09-16]https://emf.thirdlight.com/file/24/cdm30tvcyxpqsxcd10acoo2gk/towards%20the%20circular%20economy%20vol%203%3a%20accelerating%20the%20scale-up%20across%20global%20supply%20chains.pdf

[17] The blue economy, which Gunter A. Pauli proposes, is a rejection of the brown economy that recklessly consumes energy and transcends the environmentally friendly but high-cost green economy. Its characteristic is to apply the laws of the ecosystem to the economic system, advocating the cascading use of energy and nutrients based on local existing resources, creating economic value,

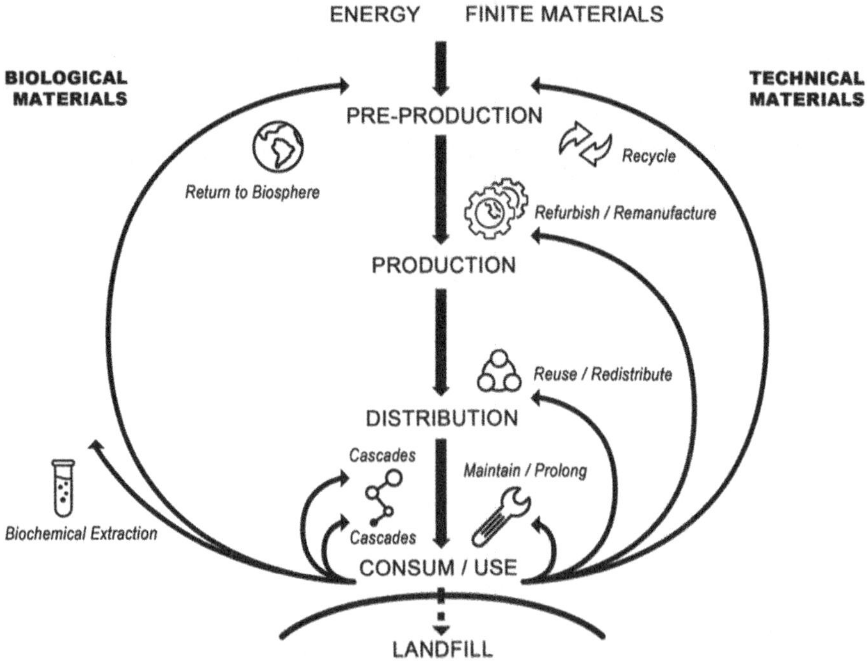

Fig. 3.2 Circular economy system model based on the lifecycle framework (Image source: https://ellenmacarthurfoundation.org)

Today, the concept of lifecycle derived from natural laws serves as a basic framework for multiple knowledge systems. Intertwined with economic and environmental considerations from its inception, it is crucial for understanding their interrelations. By employing lifecycle theory as a guideline during the design phase, we can minimize overall resource consumption and environmental impact while maximizing design value.

3.2 Product Lifecycle

A product's lifecycle includes a series of processes, from raw material extraction, production, and distribution to use and eventual waste disposal. Each of these stages has an environmental impact. Environmental impact, in this context, refers to the changes to the environment caused by human activities. This concept is straightforward: humans continually extract (input) resources (such as minerals, animals, and plants) from nature for survival and development, and in turn, discharge

while adding value to nature, achieving zero emissions, and providing an increase in job opportunities.

(output) harmful substances (such as waste, sewage, and exhaust gases) back into nature.[18] Dividing the product lifecycle into distinct stages and studying environmental impacts at each stage is essential to effectively implementing design strategies. Below, we integrate general industrial principles with the research findings of experts like Carlo Vezzoli, alongside the circular economy system model, to construct a comprehensive product lifecycle model. This model includes the five stages of a product's lifecycle: preparation, production, distribution, use, and disposal, as illustrated in Fig. 3.3. It intuitively depicts the interrelationships between these stages under ideal conditions, highlighting their input–output relationship with the environment in terms of material and energy flows, forming the basis of lifecycle design.[19]

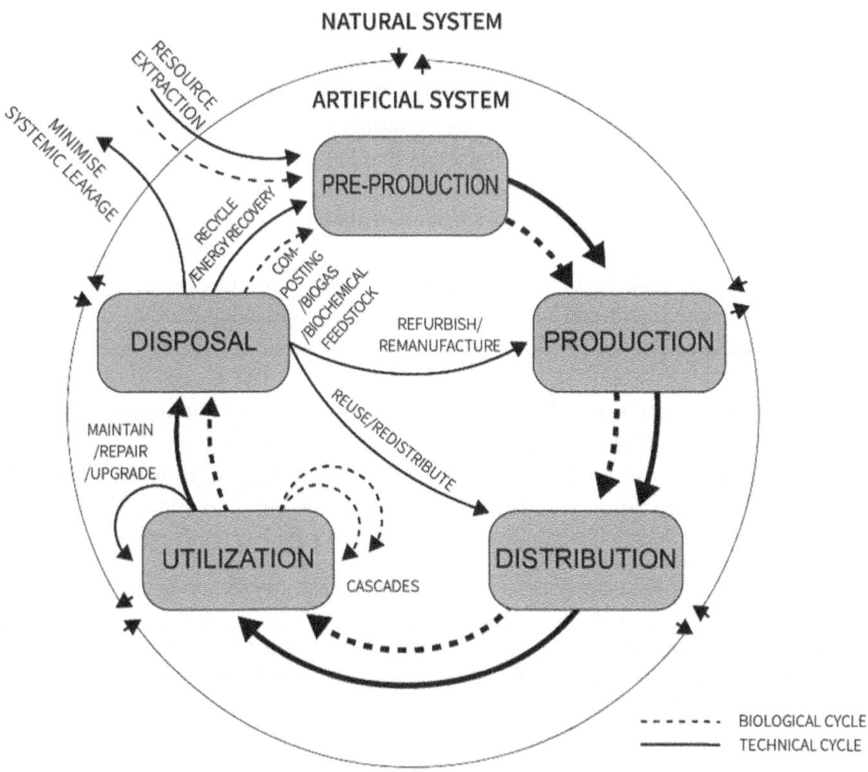

Fig. 3.3 Product Life Cycle System Model

[18] The environmental impact of harmful substances summarized are: Global warming, ozone layer depletion, eutrophication, acidification, smog, toxic substance emissions, and waste.

[19] This material classification method originates from: Vezzoli C. Design for environmental sustainability: Life cycle design of products [m]. 2nd ed. Berlin: springer,2018.

3.2.1 Production Preparation Stage

Pre-production involves preparation of the resources and semi-finished products necessary for manufacturing and includes resource acquisition, raw material, energy processing, and other foundational steps. This stage represents the starting point of the product lifecycle using both primary and secondary resources.[20]

Materials extracted from the lithosphere are predominantly non-renewable primary resources, such as petroleum, metals, and other minerals. Conversely, materials sourced from the biosphere, which typically require planting and harvesting, are renewable primary resources, such as trees and grain crops. Non-renewable mineral resources are gradually depleting. Even those intended for future generations are nearly exhausted, requiring cautious selection. Although using renewable resources presents a comparatively optimistic scenario regarding sources and applications, we still must approach exploiting them with caution. Many renewable primary resources, including various plant species like Phoebe zhennan and other high-quality hardwoods, have become extinct or are nearing extinction due to extensive deforestation by humans, thus necessitating selective future use.

Recycled resources, or secondary raw materials, originate from production waste or post-consumer waste. Research indicates that "after 300 years of predatory mining during the Industrial Revolution, over 80% of the world's industrializable mineral resources have moved from underground to above ground, accumulating around us in the form of 'garbage,' totaling up to hundreds of billions of tons and increasing by 10 billion tons each year."[21] This waste represents potential "minerals,"[22] awaiting further development and utilization. Given the increasing scarcity of primary resources, the prospects of finding applications for these secondary resources are extensive.

The pre-production stage also involves processing, storage, and transportation of raw materials. Regardless of type, each resource most go through a series of primary processing steps before it can be used in industrial production. This includes processing iron ore into steel plates, refining crude oil into plastic pellets, or

[20] Qu Yongxiang. Interpretation of "urban minerals"[j]. China nonferrous metals, 2010, 24: 30–31.

[21] Urban minerals refers to the large-scale development of recycling and reusing abandoned resources. This concept emerged in Japan in the 1980s, initially only for the recycling of rare metals in discarded household appliances. In 1988, Japanese scholar Nanjo Michio first referred to the accumulated industrial product resources on the ground as "urban minerals." (周永生,章昌平.国内外"城市矿产"研究与实践综述[j].学术论坛,2012,35(04):118–124.)N 2010, the national development and reform commission defined urban minerals as the recyclable steel, non-ferrous metals, precious metals, plastics, rubber, and other resources contained in waste mechanical and electrical equipment, wires and cables, communication tools, cars, household appliances, electronic products, metal and plastic packaging materials, and waste materials generated and accumulated during industrialization and urbanization. The quantity of these resources used is equivalent to that of primary mineral resources. Since then, the concept of urban minerals has further expanded to gradually include all discarded materials and energy.

[22] Distribution is the process of transferring goods or services from production to consumer or user by establishing sales channels.

crushing and casting recycled materials. This stage also includes transportation and storage, which contribute to product cost and can negatively affect the environment.

3.2.2 Production Stage

Production refers to the phase in which a manufacturer, guided by market targets, transforms purchased components (both new and recycled), semi-finished products, and various materials (raw or recycled) into finished products. This transformation occurs through processes such as manufacturing, assembly, debugging, and painting.

The environmental impact of production predominantly stems from energy consumption, water usage, emissions, and the generation of material waste during these processes. The steps involved in manufacturing different products are often complex and vary widely, with the result that different materials, processes, and production equipment create distinct environmental impacts. For instance, the calcination process during ceramic production consumes large quantities of energy. In the paper industry, high water consumption and the generation of significant quantities of toxic wastewater during pulping and papermaking are the main environmental concerns. The shoe industry typically consumes more raw materials during processing than other industries.

Refurbishment, reuse, and material recycling introduce unique production challenges. These include the difficulty, high cost, and carbon emissions inherent in transporting products dispersed across society to production centers. Once recovered, the materials require additional processing, including evaluation and disassembly, which further increase costs and environmental impact. To effectively balance resource conservation, cost control, and pollution mitigation, establishing a robust, efficient recycling mechanism is essential and to design products for easy disassembly and make materials simple to classify.

3.2.3 Distribution Phase

Distribution[23] bridges production and consumption. It entails packaging finished products, transporting them to various distribution units for storage, and ultimately selling them through online or offline channels, transforming products into commodities. The primary environmental impacts of distribution include the waste and pollution from packaging, as well as consumption of materials, energy, and pollution during transportation and storage.

[23] The data come from the business volume statistics the National Bureau of Statistics provides for express service companies with an annual business income of over two million yuan.

Packaging serves multiple functions: it contains and protects products, facilitates transportation and storage, minimizes potential loss during these processes, and delivers use instructions and commercial promotions. Packaging is unique in that it is both subordinate to a product and a product itself, complete with its own lifecycle of raw material processing, production, distribution, use, disposal, and environmental impact indicators. Generally, packaging has a single function and a short lifespan. If not designed or used appropriately, packaging can result in significant environmental impact and resource waste. For example, in 2019, express delivery companies in China transported approximately 63.52 billion parcels, resulting in an astronomical amount of packaging materials comprising boxes, plastic bags, and paper envelopes.[24] Beyond this, various forms of luxury overpackaging catering to vanity consumption exacerbate environmental degradation.

Transportation has a more significant environmental impact than packaging. The vast majority of products require conveyance to various storage locations and sales venues, generating substantial carbon emissions. Transportation occurs at all stages of a product's lifecycle, but its environmental impact is particularly pronounced during distribution. Common modes of transportation include trains, cars, ships, airplanes, and dedicated pipelines, each producing varying amounts of greenhouse gases. For instance, research indicates that transporting one ton of goods one mile (approximately 1.6 kilometers) by train produces 1.6 times the volume of greenhouse gases that ship transportation does, while trucks produce 10 times and airplanes 47 times that of ships, respectively.[25] Such studies are invaluable in helping practitioners, including designers, make environmentally friendly transportation choices.

3.2.4 Use Phase

Use includes utilization, consumption, the provision of services. We use some products, like cars and household appliances, while others, like detergents and instant noodles, we consume.

Products we use typically consume materials and energy and generate emissions and waste. For example, printers consume ink or toner cartridges and electricity, emitting ozone and dust. Automobiles consume fuel and emit pollutants including carbon monoxide, hydrocarbons, nitrogen oxides, sulfur dioxide, and particulate matter, collectively referred to as exhaust. Even electric vehicles, which reduce direct emissions, raise concerns regarding their sources of electricity (in China, a

[24] Dizikes P. The 6% solution [eb/ol]. (2010-11-08)[2020-03-15]. Http://news.mit.edu/2010/corporate-greenhouse-gas-1108.

[25] "Forty years of coal power transformation and new historical mission" reveals that the proportion of thermal power generation in China is gradually decreasing but still relatively high—from 80.8% in 2010 to 71% in 2017 (of which coal power is 64.7%). (王志轩.四十年煤电脱胎换骨与新历史使命[j].新能源经贸观察,2018,(09):18–22)

significant proportion still comes from coal-fired power plants[26]) and the disposal of depleted batteries.

Disposable products, such as plastics, have significant environmental impact not during use but in the production and disposal stages. Having become essential to daily life, these products have proliferated due to their large output, short use cycle, and the great difficulty of disposing of them. Their environmental impact is increasingly evident. Data from 2015 shows that only 9% of all plastic products were recycled, 12% were incinerated, and 79% were landfilled, awaiting landfill, or ended up as litter, causing severe impact to humans, animals, and ecosystems.[27] Consequently, many countries and regions, including China, have implemented plastic restriction orders to address this challenge.[28]

Many products require servicing and repairs during use. Effective servicing can significantly extend a product's lifespan and efficiency, thus reducing the overall environmental impact. Promoting sustainable consumption to raise awareness among consumers so that they maintain and repair their products is crucial to this, but as of this writing, still severely lacking in most parts of the world.

3.2.5 Disposal Phase

When we can no longer use a product and want to discard it, it requires disposal. The reasons people dispose of products are complex and not uniform.[29] While, after a long or short lifespan, a product may seem to have lost its value, this is not necessarily the case. During disposal, three primary scenarios occur: refurbishment for reuse, recycling of materials and energy, or final disposal through appropriate waste management.

The first scenario involves refurbishing the usable parts of discarded products, making them look new and functional again and encouraging their reuse or repurposing for different functions (see Chap. 5). For instance, Apple recycles products through various channels, refurbishes them, and sells them in its online Certified Refurbished Products store. In 2018, Apple refurbished more than 7.8 million

---

[26] Quoted from the adaptation of Geyer et al. (2017) and Jambeck et al. (2015) in the 2018 report single-use plastics: A roadmap for sustainability by the United Nations Environment Programme.

[27] The European commission passed a proposal to ban single-use plastic products in 2018; China issued the "opinions on further strengthening plastic pollution control" (2020) based on the "notice on restricting the production, sale and use of plastic shopping bags" (2007), and formulated clear delisting plans for several types of plastic products such as non-degradable plastic bags, disposable plastic tableware, disposable plastic products in hotels, and express plastic packaging.

[28] For the reasons why people discard products, see sect. 6.1.

[29] See detailed data in apple inc.'s "environmental responsibility report: 2019 progress report—a comprehensive review of fiscal year 2018."

devices and recycled over 48,000 tons of electronic waste.[30] However, refurbishing and reusing also consume resources and emit pollution, causing new environmental impacts. Some discarded products, in good condition, may only require simple cleaning for reuse, while upcycling can transform old glass bottles into decorative items or lamps (see Chap. 5).

The second scenario occurs when it is no longer possible to refurbish a product, but its materials are recyclable, compostable, or incinerable to generate energy (see Chap. 4). There are technologies to recycle, clean, melt down the majority of metals or make them into secondary raw materials, constituting an urban mining phase in production preparation. We can re-pelletize some plastics for reuse. Typically, the quality of recycled materials deteriorates downcycling until they become waste. Therefore, it is crucial to maximize their use within limited cycles. We can compost biodegradable materials, such as starch-based agricultural films and kitchen waste, into bio-fertilizers. We can burn other forms of waste as fuel to generate electricity, contributing their final value to recovering energy. In some cases, incineration effectively eliminates non-recyclable harmful materials, such as with medical waste. Although incineration technology has advanced, making pollutant emissions negligible, it discounts the value of materials and indirectly leads to resource waste.[31] This is not a criticism of the technology itself but reflection on our current lifestyles and consumption habits.

The third scenario involves dumping waste in landfills, which ideally should not occur but seems unavoidable under the current circumstances. Dumping can have extremely negative effects on the surrounding environment, including pollution from improper waste handling and disposal. Many cities face the problem of being surrounded by waste. Even with effective separation of recyclable and combustible materials, some waste, including bricks, ceramics, and slag, remains. Theoretically, post-incineration ash could have a new life as a building material, but the cost is often too high for businesses to absorb. Therefore, the best outcome is to manage waste as effectively as possible, including establishing technically equipped landfills, legally disposing of waste, and treating waste in ways that does not harm the surroundings.

All stages of a product's lifecycle are interconnected and mutually influential. From initial material selection to production to use, each stage directly impacts the form of a product's final disposal. Problems at one stage often result from inadequate planning and design at earlier ones. The role of design is not to achieve technological breakthroughs, but instead to intervene at the source and affect the entire lifecycle of a product, ultimately facilitating the transformation of our production and consumption methods.

---

[30] According to national statistics, about 228 million tons of domestic waste were collected in China in 2018, of which about 102 million tons were incinerated. Are there usable resources in this incinerated waste?

[31] Liu xin. Concept, development and practice of sustainable design[j]. Creativity and design, 2010(2): 36–39.

The above division into stages emerges from traditional industrial rules. With the rapid development of new technologies like artificial intelligence and 3D printing, activities at different stages may undergo new integrations and intersections. Some traditional problems may persist or gradually diminish, while new ones will arise. Influenced by human concepts and desires, artificial intelligence may eventually surpass human control, presenting challenges beyond the scope of this chapter. If 3D printing achieves remote data transmission and local manufacturing, it could significantly reduce the environmental impact of transportation and packaging (see Chapter 11). However, 3D printing materials and resource consumption still pose challenges. The convenience of manufacturing and the ability to meet individual needs may further stimulate depletion because individuals consume more resources than before. For the foreseeable future, these traditional stages remain the mainstream view of lifecycle research, aiding design practitioners in understanding environmental impacts at different stages of production and identifying design opportunities.

3.3 Lifecycle Design Strategy

From the above analysis, it is clear that design can intervene at all stages of a product's lifecycle and play a positive role in reducing environmental impact. This is precisely where the value of lifecycle design lies.

3.3.1 System Thinking

The core feature of lifecycle design is to shift from focusing solely on designing a particular product to considering the environmental impact of the entire process associated with that product. This holistic, system-based approach helps prevent local improvements from leading to misrepresentation and allows for a comprehensive understanding from a broad perspective.

Some green designs, albeit well-intentioned, not only fail to reduce products' environmental impact, but increase them. Design can sometimes cause great damage to nature under the guise of environmental protection. For instance, consider cardboard furniture. People generally regard paper products as the most natural materials, since they boast recyclability, degradability, ease of transportation and storage, and low cost, aligning well with green and environmental protection concepts. Compared to traditional furniture, cardboard furniture seems to have clear advantages in terms of production and distribution (Fig. 3.4). However, the lifespan of cardboard furniture during use is quite limited, which necessitates frequent replacements. In terms of durability, it obviously cannot compare with traditional furniture lacking any explicit environmental protection design. Additionally, the recycling, transportation, and reprocessing of discarded cardboard involve

Fig. 3.4 The cardboard chair is not greener than a traditional chair built without any considerations for the environment

significant energy consumption and emissions. Recycling paper requires large quantities of water, generates pollution, and consumes energy. Therefore, a system analysis from the perspective of its entire lifecycle reveals that cardboard furniture actually incurs a higher environmental cost than its conventional counterparts.[32]

Intervening at a system level might seem beyond design's sphere of influence. Whether designers have sufficient capability and influence to undertake this critical task is uncertain, but there are nonetheless an increasing number of instances of systemic design (see Chap. 12). Contemporary society has broadly accepted the need for sustainable development, and many companies' development strategies integrate environmental awareness, which has even become part of branding. Thus, lifecycle design has extended well beyond the design department to become a shared mission across management, marketing, engineering, and other fields. As noted at the beginning of this chapter, lifecycle design may not only protect the environment, but it can also prevent economic losses associated with end-of-life treatment, which benefits companies aiming to engage in long-term, sustainable development.

3.3.2 Functional Unit

A fundamental concept in lifecycle design is that of the functional unit. What we design is not just a product, but also a sequence of processes aimed at fulfilling a specific function. *Consequently, when evaluating the environmental impact of a design, we cannot confine our analysis solely to the product; rather, we must assess the environmental ramifications of products, services, and processes that fulfill an equivalent function.*[33] The functional unit serves to quantify a desired product

[32] Vezzoli C. Design for environmental sustainability: Life cycle design of products[m]. 2nd ed. Berlin: Springer, 2018: 45.

[33] National environmental management standardization technical committee. Principles and framework of environmental management life cycle assessment: gb/t 24,040—2008 [s]. Beijing: China standards publishing house, 2008:1.

function.[34] Ensuring that we evaluate different systems according to a common measurement facilitates meaningful comparisons. To illustrate, let us compare buses and automobiles: throughout their separate lifecycles, buses tend to have a significantly larger environmental footprint than cars. They require more materials and resources for production, involve more complex manufacturing processes, consume more energy during operation, and generate more waste than cars. However, when we evaluate their environmental impact in terms of fulfilling an equivalent function, the results change markedly. We take the number of kilometers per person as the functional unit and assume that the average car carries two people while a small bus carries 20. If both vehicles transport passengers the same distance, then a bus has the carrying capacity of 10 cars. Clearly, based on current data calculations, when viewed through the lens of functional requirements, buses emerge as significantly more environmentally friendly than automobiles.

The functional unit serves to quantify a product's overall performance, allowing us to make environmental assessments that transcend individual products and observe the core of functional requirements. This approach to design aims to deliver satisfaction and pleasure to users through a product's function, rather than the product itself. Although questions of a product's aesthetic and emotional attributes are important because they hold considerable value for users, these fall outside the scope of this chapter.

3.3.3 Design Strategies

Breaking the rules may generate superior outcomes, but knowing them still matters. Evaluating a product or service hinges on more than just minimizing environmental impact or potential ecological benefits; it encompasses a myriad of other factors. As noted above, effective design must meet criteria that span functionality, usability, technology, economy, regulations, culture, and aesthetics, thus providing a comprehensive solution to people's needs. In this light, we propose the following lifestyle-design inspired strategies.[35]

1. *Minimize Material and Energy Consumption, Reduce Harmful Emissions:* This overarching goal pervades all lifecycle stages and constitutes the primary strategy. Amid the unprecedented proliferation of material products, curtailing resource consumption and environmental pollution is imperative to sustainable

[34] The guidelines mainly refer to Carlo Arnaldo Vezzoli's (2018) *Design for environmental sustainability: Life cycle design of products* (second edition) and the U.S. Environmental Protection Agency's *Life cycle design guidance manual* (https://cfpub.epa.gov/si/si_public_record_report.cfm?lab=ord&direntryid=35524)

[35] Non-renewable resources, such as minerals and fossil fuels, although derived from nature, recover or replenish at a very slow rate and cannot meet current rates of consumption. Although some of their by-products, such as plastics and metals, can be recycled multiple times, they are easily discarded and thus still considered non-renewable resources.

development. Any meticulously crafted design strategy that fails to address this imperative risks being ineffective. Further elaboration on this theme occurs in Chap. 7.

2. *Privilege Renewable, Biocompatible, and Non-Toxic Resources:* This strategy is inextricably linked with minimization. The escalating scarcity of non-renewable resources[36] requires prioritizing renewable[37] raw resources and secondary materials when preparing for production to mitigate resource depletion. Using biodegradable materials presents novel design prospects, while curbing pollutant emissions. However, designers must take account of the harsh degradation conditions associated with certain degradable materials. Designers, moreover, should avoid toxic materials in their selection, avoiding them or placing them where they are difficult to access and easy to recycle. Harnessing renewable energy can greatly reduce emissions and aligns with energy security considerations throughout all lifecycle stages (see Chap. 7).
3. *Optimize Product Lifespan:* Optimizing products' lifespan entails designing for reusability, durability, or frequent utilization. This approach not only reduces resource consumption by obviating the need for replacement products, but also curtails waste generation. Nonetheless, extending product lifespan is contingent upon ensuring product safety and reliability. Research on select electrical and electronic products suggests an optimal service life of approximately 20 years for refrigerators with relatively mature and stable technology, contrasted with 7 years for laptops, reflecting a dynamic market prioritizing resource utilization and business feasibility.[38] Increasing product use frequency entails intensive utilization during a limited lifecycle to maximize utility and value, exemplified by shared products and corresponding product-service system designs (see Chaps. 5, 6, and 9 for further elaboration).
4. *Prolong Material Utilization:* Recycling and composting give new value to materials rendered unusable, thereby reshaping product disposal methods and providing fresh options for production by supplying renewable secondary materials (see Chap. 4 for details). However, these processes engender new challenges including difficulty recycling, energy consumption, financial outlay, and degradation of material quality. Consequently, the focus of design should center on reducing material use, using materials optimally, and planning material end-of-life scenarios.

[36] Renewable and non-renewable resources together constitute the resources available to humans. Renewable resources include timber, plants, microorganisms, as well as solar energy, wind energy, and biomass energy.

[37] Bakkerc, Wang F, Huisman J, et al. Products that go round: Exploring product life extension through design [j]. Journal of cleaner production, 2014, 69: 10–16.

[38] Chiu M C, Okudan G E. Evolution of design for x tools applicable to design stages: A literature review[c/ol]//ASME 2010 international design engineering technical conferences and computers and information in engineering. ASME, 2010: 171–182. [2021-09-16]. https://doi.org/10.1115/detc2010-29091

5. *Champion Disassembly Design:* Design for disassembly has importance during use and disposal in terms of facilitating repair, part replacement, and refurbishment.[39] It mitigates recycling costs, increases resource utilization rates, and bolsters product lifespan and material utilization duration. Incorporating disassembly design for vulnerable, easily soiled, and upgradeable parts facilitates product maintenance, upgrades and updates, thereby extending product longevity. Considering the disassemblability of diverse materials promotes material classification, recycling, and circular regeneration (refer to Chap. 4 for comprehensive discussion).

3.4 Optional Strategy Set

The basic principle of lifecycle design is to transcend the end-of-life-treatment mindset, go back to the source, and focus on the design of systems, wholes, and processes. The design strategies outlined above can significantly reduce the environmental impact of products and services. However, in practice, applying them does not ensure the achievement of sustainability goals. Universal solutions that claim to address all problems are often questionable. Simply applying well-established design guidelines cannot guarantee problem resolution and may sometimes increase negative environmental impacts (e.g., the example of cardboard furniture). Although system thinking and functional unit analysis can help us reduce errors, the significant differences between various types of products necessitate targeted analysis in order to weigh various measures and implement the most promising.

Different products—cars, furniture, or disposable paper cups—will engender different strategy priorities. For instance, the environmental impact of furniture products is mainly at the raw material acquisition and production stages, with some impact during disposal, and negligible impact during use. Therefore, extending product lifespan through durable design is the primary strategy for furniture. For automobiles, energy consumption and pollution during use are the most significant problems. Thus, minimizing these impacts through clean energy use is crucial. Additionally, promoting product-service systems for car-sharing and car-leasing is important. For disposable paper cups, the primary issues are to resource the environmental pollution large waste volumes cause. Disposable cups often require coatings for waterproofing,[40] making effective recycling or natural degradation impossible. Thus, using biodegradable materials for disposable products may be the

[39] The inner coating of paper cups is mostly made of polyethylene, which is extremely difficult to separate from the paper material, making such paper cups non-recyclable.

[40] The Ellen MacArthur foundation proposes three principles for the circular economy: To avoid waste and pollution from the outset of design (design out waste and pollution); to extend the use cycle of products and materials (keep products and materials in use); and to promote the regeneration of natural systems (regenerate natural systems).

best strategy, provided new material technologies are reliably applied and cost-effective.

The aim of lifecycle design is to both reduce environmental impact and account for economic dimensions, such as the cost and benefits of adopting certain strategies. The Ellen MacArthur Foundation highlights the need to avoid waste and pollution from the initial steps of a design as a fundamental principle of the circular economy.[41] However, current research indicates that lifecycle design does not address social dimensions. Since design inherently includes insight into human nature, capturing social trends and shaping consumer culture,[42] design for sustainability should integrate various social perspectives and methods. Despite this, lifecycle design remains one of the most comprehensive, systematic approaches in sustainable design.

[41] Hauschild M, Wenzel H, Alting L. Life cycle design-a route to the sustainable industrial culture [j]. CIRP Annals, 1999, 48(1): 393–396.

Chapter 4
Design for Recycling

Abstract Recycling involves transforming a product's constituent materials into secondary raw materials after the product has reached the end of its life. This process effectively extends the lifecycle of material resources, conserving resources and protecting the environment.

Keywords Recycling · Product lifecycle

Ideally, in future, ours will be a closed-loop[1] material circulation process, where we reuse or recycle all items and materials without waste, akin to cycles in nature, which use and continually cycle everything. However, we are still very far from that ideal. Currently, incineration and landfill are the ultimate fate of most materials, which not only wastes resources but also pollutes the environment and occupies land. As noted in Chap. 3, although incineration can realize partial energy recovery and advanced technologies can minimize pollution emissions, incineration represents a net loss when considering all the resources and energy a product consumes during its lifecycle. Designers, collaborating with engineers, entrepreneurs, policymakers, and consumers, may be able to extend the life of products and materials as much as possible.

There are several options for what to do with waste at the end of its lifecycle: organic waste, such as food residues, fruits, and vegetables, and fallen leaves, is treatable with biotechnologies to produce fuel, fertilizer, packaging, or decorative materials. Industrial product waste, such as unusable household appliances, mobile phones, or equipment parts, contains materials like aluminum, gold, copper, and tin that we can break down and recycle to create new materials and products (see Fig. 3.2).

Future trends in material development point to the widespread use of biomass degradable materials in industrial products (such as PLA, PHA, PBS),[2] which could

[1] The closed and open loop refer to the relationship between material flow and systems, where closed loops refer to materials circulating mostly only within a single system; the corresponding open loop connotes flows of material that involve multiple systems.

[2] PLA is short for polylactic acid, PH is short for polyhydroxyalkanoates, and PBS is short for polybutylene succinate. Currently, the large-scale use of bioplastics still faces issues in terms of

X. Liu et al., *Design for Sustainability*, Sustainability and Interdisciplinarity,
https://doi.org/10.1007/978-981-96-8657-5_4

significantly reduce the environmental impact of waste. These materials offer limitless prospects as proactive alternatives. However, in the foreseeable future, traditional industrial materials (metals, plastics, rubber, ceramics, glass, and various composites) will remain the primary raw materials for product manufacturing. Therefore, extending the lifecycle of materials and promoting recycling are crucial strategies in the introduction of sustainable design.

4.1 What Is Recycling

Recycling is the process of collecting, dismantling, and cleaning discarded products or materials, then remaking them into new materials and products. In other words, when a product reaches its end of life, its constituent materials can be reborn from the ashes, transformed into secondary raw materials, and used to create new products. For example, scrapped cars contain significant quantities of steel, non-ferrous metals, plastics, rubber, glass, and textiles. More than 90% of the steel and non-ferrous metal parts are recyclable, as is over 50% of glass and plastics. From a scrapped car, we can recover around 1000 kilograms of waste steel and approximately 50 kilograms of non-ferrous metals.[3] Once effectively recycled, these valuable materials can enter the process of recycling to manufacture new products.

The Tokyo Olympics put the philosophy of resource recycling into practice in an exemplary way. According to the Tokyo Olympic Organizing Committee, Japan launched a campaign to collect discarded mobile phones and home appliances on April 1, 2017. Over 2 years, the committee collected approximately 78,985 tons of small appliances and 6.21 million old mobile phones nationwide. After extraction these yielded about 32 kilograms of pure gold, 3500 kilograms of pure silver, and 2200 kilograms of copper. These recycled metals generated some 5000 medals, with ribbons made from recycled textile fibers.[4] This marked the first time in Olympic and Paralympic history that citizens participated in the medal-making process through waste recycling. The significance of this initiative extends beyond the recycling large quantities of metals. It also highlights the importance of social participation and global environmental awareness. The process made people who had previously been unaware that there are valuable resources hidden in our everyday waste.

Typically, the environmental impact of regenerated materials is significantly less than that of producing the same quantities of raw materials. Although the process of recycling—including collection, transportation, disassembly, cleaning, and processing—also consumes energy and emits pollutants, statistical data demonstrate

---

commercialization and performance, while their conditional degradability require a good classification and recycling system. https://www.european-bioplastics.org

[3] Tian Guangdong, Jia Hongfei, Chu Jiangwei. Theory and Practice of Automobile Recycling [M]. Beijing: Science Press, 2016: 5.

[4] See the official website of the Tokyo Olympics https://tokyo2020.org/

that recycled materials have clear advantages in terms of both economic and environmental benefits. The efficiency of recycling industrial waste or tailings produced during manufacturing is higher than recycling post-consumer waste due to the relatively homogenous, concentrated, and easily identifiable nature of the materials and may even allow a product's lifecycle to complete itself during production.[5]

The physical and mechanical properties of recycled materials are generally inferior to those of raw materials, due to a process called downcycling. Some metals, however, constitute exceptions; if the recycling process is well managed, it is possible to fully maintain the quality of recycled metals. If the structure of the raw materials is damaged, or if impurities, pollutants, or coatings are mixed into the material, or it is exposed to ultraviolet light for extended periods, this can result in the quality of the recycled material being compromised, affecting its future applications. Most materials are not indefinitely recyclable; with each regeneration process, they degrade until they are no longer usable. At that point, incineration for energy recovery may be the best option.

Several factors affect a material's recyclability, such as mixing with other materials or embedding in products that are difficult to disassemble. For example, shirts made from a 60% cotton and 40% synthetic polyester blend cannot enter the current recycling system but must be landfilled or incinerated. Similarly, disposable paper cups, which appear to be made of paper but have a waterproof plastic film inside, are not processable as either paper or plastic, posing a recycling challenge.

Difficult-to-disassemble products present significant obstacles to recycling. Until the 1980s, automobile assembly plants mainly used adhesives to attach interior panels, such as plywood and synthetic fabric, to the car body's steel plate. While this method was efficient, it created problems, including health hazards to workers due to volatile adhesives. Workers dismantling scrapped automobiles had to spend considerable time and effort separating materials and the quality of recycled materials was often low. Today, automobile manufacturers use plastic snaps, ensuring worker safety and increasing the convenience and efficiency of disassembly. Designing for disassembly is crucial to effective recycling.[6]

4.2 Steps of Recycling

Recycled materials must usually pass through the following steps: collection and transportation, identification and classification, disassembly or shredding, cleaning, or washing, and secondary raw material processing, as shown in Figure 4.1.[7]

[5] Vezzoli, C., *Design for environmental sustainability: Life cycle design of products* [M]. second ed. Berlin: Springer, 2018: 156.

[6] Disassembly design is a design philosophy that makes the components and materials of a product easy to separate by pre-design, with the aim of improving the maintainability of products and the convenience of classification during recycling.

[7] Same as 5: 157

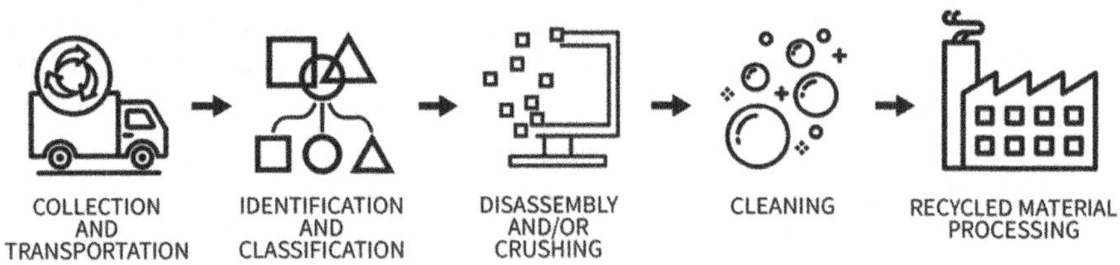

Fig. 4.1 Recycling Process Diagram

Collection and Transportation: The initial step in the recycling process involves collecting and transporting discarded items to factories or warehouses. This stage involves various participants, with the user playing a pivotal role. User habits and behaviors determine when a product reaches the its end of life and begins the recycling journey. Effective, proactive waste sorting by users can significantly enhance recycling efficiency and reduce costs. Transportation of wastes also incurs economic costs and environmental impacts, often underestimated, likewise influencing recycling efficiency.

Identification and Classification: The first task once materials arrive at a storage facility is identifying and classifying those which are recyclable. The collection process is typically imprecise, entailing transportation of entire products or parts thereof. Therefore, clearly marking material types during the design phase—using standards like the Chinese national Packaging Recycling Mark (GB/T 18455—2010)—aids workers in quickly identifying and sorting materials.

Disassembly/Shredding: Disassembly and shredding facilitate final material separation. Disassembly involves separating materials, such as metals from plastics or different types of thermoplastics. We can recycle some composite materials, known as compatible materials, without compromising product quality. Components that are dismantlable have higher economic value than while complex products that are not, which undergo shredding, with materials separated using magnetic separation, induction, and suspension techniques. The ease of disassembly is a crucial factor in recycling, one which designers can significantly influence.

Cleaning: After separation, many materials remain contaminated with dirt, labels, adhesives, and ink residues, requiring cleaning or washing to ensure they are useable. Design strategies should consider how to avoid such contaminations, for instance, by using molded text or graphics instead of ink printing.

Secondary Raw Material Processing: To ensure the quality of secondary raw materials often requires special processes or additives are often employed. While this typically does not affect the use of recycled products, it can negatively impact subsequent recycling cycles. For example, plasticizers used to enhance polymer material performance can reduce the quality of recycled materials, pollute the environment, and pose health risks.

Ensuring material quality often requires implementing special processes or adding chemical substances when manufacturing secondary raw materials. In many cases, this does not affect the use of recycled products, but it has a negative impact when we recycle object multiple times, such as the plasticizers added to improve the performance of polymer materials, which not only reduces the quality of recycled materials, but also pollutes the environment and affects human health.

Broadly speaking, the disposal of organic waste with the goal of resource reuse also constitutes recycling. In most cases, organic waste must also go through collection, storage and transportation, and treatment. When non-degradable impurities become mixed in during collection, it significantly affects the treatment of organic waste. Thus, promoting waste sorting and public science education are essential. In addition, we must devote attention to pollution and the NIMBY (Not In My Backyard) effect[8] during storage and transportation. Treatment of organic waste can yield various resources, including animal feed, organic fertilizer, biomass energy (e.g., biogas), and biomass oil (biodiesel).

4.3 Design Principles of Recycling

As noted above, the design stage determines over 80% of a product's future environmental impact.[9] If designers lack understanding and do not intervene during the engineering phase, dismantling and classifying recycled products becomes extremely difficult once the production and sales have commenced. Therefore, during the initial design phase, it is extremely important to adhere to principles such as selecting the appropriate materials, reducing the number of types of materials, ensuring easy identification, designing for cascading uses, and facilitating collection, storage, cleaning, and disassembly to ensure effective recycling.

Material Selection: The variety of materials available is extensive. Some are easy to recycle and may undergo multiple recycling rounds. These include steel, aluminum, glass, plastic, paper, and even concrete. Other materials are nearly impossible to regenerate or require complex, costly processes to do so. For example, ceramics, coated glass, coated paper, ultra-thin plastic packaging, and medical waste can currently only be incinerated or dumped in a landfill due to technological limitations.

Compared to other recyclable materials, aluminum performs exceptionally well (Fig. 4.2). Aluminum exhibits nearly infinite recyclability and high durability. According to data from the Aluminum Association of the United States, humans have produced approximately one billion tons of aluminum, with about 75% still in

---

[8] Tang Honglin defines the NIMBY effect in "Analysis of the Root Causes and Countermeasures of the 'NIMBY Effect'" as "local residents taking strong, emotional group resistance actions due to concerns about the adverse effects of surrounding construction projects on the surrounding environmental quality and physical health." See: http://www.Rmlt.com.cn/2020/1104/597831.shtml

[9] European Union. Ecodesign your future: how ecodesign can help the environment by making products smarter [r/ol]. (2014-11-24). [2021-02-08]. https://op.europa.eu/en/publication-detail/-/publication/4d42d597-4f92-4498-8e1d-857cc157e6db

Fig. 4.2 Common aluminum products (Image source: https://www.conserve-energy-future.com)

use, and some aluminum recycled an incalculable number of times.[10] Despite this, the environmental impact of raw aluminum mining and refining is significantly high compared to those of steel. Therefore, using products made from recycled aluminum is the environmentally friendly choice.

With the development of the chemical industry, plastic products have become ubiquitous. Our daily dependence on plastic is beyond imagination. The significant impact of plastic waste, including microplastics,[11] on the environment has led to growing doubts and concerns about the efficacy of this organic polymer material. Some potential hazards are still unclear, prompting further apprehension. It is worth contemplating whether these issues are inherent to plastic itself or a result of human misuse.[12]

---

[10] Data sourced from the Aluminum Association website: https://www.aluminum.org/aluminum-advantage/facts-glance

[11] The National Geographic website defines microplastics as tiny plastic particles from industrial products and the decomposition of large plastic products, with a diameter of less than five millimeters. As a pollutant, microplastics pose certain risks to the environment and animal health. (https://www.nationalgeographic.org/encyclopedia/microplastics/)

[12] Plastic originates from the fossil fuel petroleum. Originally invented in the early twentieth century as a by-product of the petrochemical industry, high-quality, inexpensive plastic products quickly pervaded our lives due to technological breakthroughs. They became ubiquitous and the primary material for disposable items. World Economic Forum data shows that 32% of plastic packaging ends up as litter, 40% is landfilled, 14% incinerated, and 14% recycled. However, only 2% undergoes effective recycling. Littering, landfilling, and incinerating plastic not only destroys the environment, but also has adverse effects on human and animals' biological health. In future, formulating new cross-regional policies, laws, and market rules will place responsibility for plastic pollution on producers and their processing methods, forcing companies to research and use degradable alternative materials. Universal education and popular science will influence consumption concepts, achieving changes at the source. Therefore, only with unified goals and global cooperation can we solve the thorny plastic problem and protect our human home. Source: the documentary *The Story of Plastic* (2019) produced by Pale Blue Dot Media, directed by Deia Schlosberg.

From a recycling perspective, thermoplastics, which account for more than 80% of all polymer materials, have good recycling potential. This is especially true for the most common types, such as PET (Polyethylene Terephthalate) and HDPE (High-Density Polyethylene). While thermosetting plastics are not recyclable through liquidification at high temperatures, they can be crushed and added to thermoplastic materials as auxiliary components (Fig. 4.3).

People often consider paper the most recyclable material (Fig. 4.4), but this does not inherently make it an environmentally friendly choice, as the example of the cardboard chair illustrates. There are many types of paper materials and their states during recycling vary. For instance, corrugated cardboard, packing boxes, and newspapers are valuable recyclables, whereas recycling toilet paper, napkins, laminated cardboard, paper cups, and contaminated paper is difficult. The fundamental purpose of paper recycling is to protect trees and forests and to reduce landfill waste.

However, paper's light, scattered nature and its varied uses mean that ineffective sorting and recycling at the front end can result in it becoming mixed in with other garbage and non-recyclable due to secondary pollution. Moreover, papermaking processes—whether of virgin or recycled paper—have a significant environmental

Fig. 4.3 Recyclable plastic products. (Image source: https://www.eco-ricicli.it)

Fig. 4.4 Recyclable paper products (Picture source: https://www.aparasdepapelsaojorge.com.br)

impact. Therefore, designers should carefully consider the environmental implications before opting to use paper materials.

Discarded concrete is easy to recycle, despite what most people may think. The quantity of construction waste, including concrete, is astonishingly high, accounting for about 25% of all solid waste globally.[13] In China, the annual production of construction waste exceeds 1.5 billion tons, constituting about 40% of urban waste.[14] Construction waste with extremely low calorific value is obviously not incinerable. If dumped in a landfill, it not only occupies land and pollutes the environment, but also increases transportation costs. How to make good use of this huge resource? In fact, there are many uses for recycled concrete, such as preparing coarse or fine aggregates for the manufacture of recycled bricks and other environmentally friendly building materials, or using it in construction, which can not only reduce environmental impact but also reduce economic costs (Fig. 4.5).

When designing, practitioners should not only consider whether certain materials are suitable for recycling, but also conduct comprehensive analyses based on a material's entire lifecycle, which may involve sourcing raw materials (transportation costs and carbon emissions), the environmental impact of manufacturing processes, and the destination and use of the material once recycled.

Select Appropriate Materials: From both an economic standpoint and in terms of mitigating environmental impacts, opting to minimize the array of materials in

[13] Leblanc, R. The importance of concrete recycling [EB/OL]. (2017-09-29)[2021-02-08]. https://www.thebalancesmb.com/the-importance-of-concrete-recycling-2877756

[14] China Building Waste Resource Industry Technology Innovation Strategic Alliance. China Building Waste Resource Industry Development Report (2014)[R]. China Commerce Industry Research Institute, 2015.

Fig. 4.5 Concrete construction waste and aggregated bricks made from waste (Picture source: https://www.thebalancesmb.com)

products or their components is a logical choice. In certain instances, market demands or perceived aesthetic preferences may drive designers to develop intricate embellishments or use multiple materials. However, this notion is antiquated. Embracing simplicity and moderation not only is stylistically eloquent. It also significantly influences material selection and use.

Indeed, from a recycling standpoint, products fashioned from a singular material prove simpler to identify, categorize, and process—whether basic drinking vessels or multifaceted products. Moreover, items crafted from a singular material can still exhibit a diverse range of forms and effectively fulfill various functional objectives. A prime example is the Celle office chair. Conceived by Jerome Caruso for Herman Miller, it comprises compact modules in a cellular structure capable of accommodating individual preferences for support firmness and elasticity, thus ensuring optimal comfort. The entire piece is fashionable from a single polymer material, facilitating easy disassembly and separation to ensure effective recycling.

Ensure Material Identification Simplicity: In general, users, primary recyclers (scavengers), and sorters rely on sensory experiences such as visual, tactile, and auditory cues to discern material types and effectively distinguishing basic categories like metal, glass, and plastic. However, in the recycling industry, precise classification of recycled materials is imperative, particularly for complex products or materials of uncertain origin. Thus, identifying an object's material composition assumes paramount importance.

Marking the properties of materials, including factors such as lifespan, recycling potential, and additives, thus becomes a fundamental design principle. Such markings should be prominently visible on components and adhere to a standardized identification system. Designers should give special consideration to toxic and harmful materials. This design principle is straightforward and well-defined. For instance, the Plastic Material Coding system established by the American Plastics Industry Association and the ISO 11469:2016 standard (General Identification and Marking of Plastic Products), issued by the International Organization for

Standardization, both contain clear regulations (see Fig. 4.6) and are widely adopted by product designers.

Adopt a Cascade Design Approach: As its name suggests, the cascade approach involves pre-designing a material recycling pathway such that materials transition from one product or component to the next, meeting progressively lower performance requirements until they reach a point where reuse is no longer viable and undergo energy recovery via incineration. A comprehensive cascade approach design ensures maximal extraction of material value and potential, effectively leveraging scarce resources. For instance, food-grade plastics endure successive downgrading each time they are recycled. Many public facility materials, such as plastic guardrails, isolation piers, public seats, and trash cans, use materials recycled multiple times. This approach has widespread application in forestry and the wood product industry, where materials progress from solid wood to veneer, wood products (such as furniture or building materials), particle board, fiberboard, and bio-based chemical products, ultimately yielding thermal energy through incineration.[15] While maintaining product functionality, the cascade approach optimizes material lifecycle extension, yielding significant economic benefits.

Optimize for Collection, Storage, and Transport: Efficient collection and storage hinge on both product and service system design principles. First, in product design, maximizing standardization and regularization minimizes space requirements during stacking for storage and transportation. Prioritizing reductions of product weight and volume over extravagant design features (e.g., unique shapes or flashy packaging) facilitates collection and storage. Bulky, heavy products pose challenges during collection and transportation; abandoned mattresses and furniture by community trash bins often linger for days due to the logistical difficulties of moving them. In contrast, foldable cardboard boxes and lightweight plastic bottles are easier to collect. Evian Company enhances transportation efficiency by designing mineral water bottles with compressible patterns, optimizing storage space post-disposal.

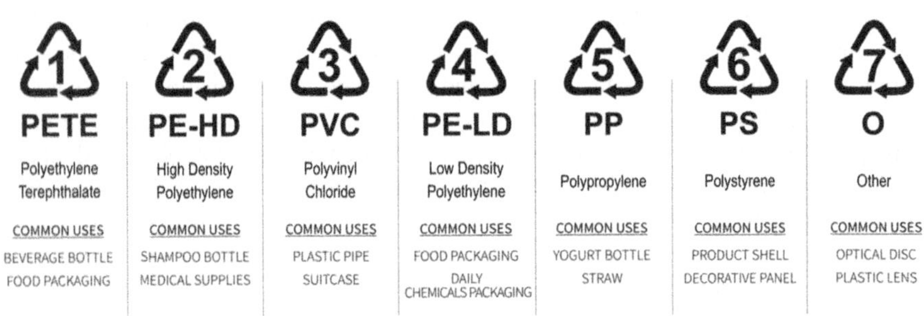

Fig. 4.6 General identification marks for several types of plastic products (Based on the American Plastics Industry Association's Plastic Material Coding)

[15] See: https://www.ceguide.org/strategies-and-examples/dispose/cascading

Furthermore, modular and easily dismantled designs offer additional solutions to this issue.[16]

On the flip side, designers possess the ability to propose targeted recycling system designs, contingent upon a thorough understanding of existing waste-recycling infrastructures. The targets for such designs include government-operated entities, commercial institutions, and individual practitioners, which should delineate the materials they handle and their operational methodologies. Contemporary, professional recycling platforms catering to various items like mobile phones, appliances, and clothing notably expedite waste collection, enhancing user convenience and improving the retrieval of valuables from waste streams in an expedient, efficient manner. Designers have ample opportunities to contribute further, such as developing educational materials on waste classifications and recycling methodologies.

Emphasize Cleanability: Cleaning considerations extend beyond post-recycling surface decontamination; they must be integral to the design process. Key principles include minimizing surface embellishments and unnecessary coatings, such as decorative electroplating, while maintaining functionality. Eschewing surface printing,[17] particularly on inner surfaces, is advisable, employing alternative methods instead like plastic, printed films for essential labels and product information, and easily detachable nameplates. Coloring raw materials rather than painting their surfaces not only streamlines paint removal from recycled parts, but also augments durability and cost-efficiency. Embedding product identification information directly onto the product through molding serves to obviate the need for printing and separate nameplates. These fundamental principles should become subconscious knowledge among product designers.

Ensure Ease of Disassembly: Ensuring easy disassembly is crucial for efficiently and effectively separating incompatible materials, facilitating processes such as identification, sorting, cleaning, and material processing. This principle not only reduces economic costs, but also plays a vital role in repairs, upgrading, and reuse of products.

In some cases, for reasons of product safety, operational stability, and intellectual property protection, companies do not want users to disassemble products themselves, so they intentionally design them to increase the difficulty of disassembly, sometimes necessitating destructive methods for disassembly. While this concern is understandable, companies must also consider the need for repairs and the proper disposal of products at the end of their lifecycles. For these reasons, we recommend using special tools or equipment for disassembly operations.

The focus of easy-to-disassemble design is minimizing the time (and consequently the cost) of disassembly, improving efficiency, and predetermining the

[16] Evian designed its water bottle with a spiral line on the body, which not only facilitates drinkers holding it in their hands, but also instructs users to compress the empty bottle to the smallest possible volume after drinking, to minimize the space occupied during transportation.

[17] Some transparent glass and plastic products manufacturers, to avoid printing information on the outer surface of the product, adopt the method of printing on the inner surface. This creates great difficulty when cleaning recycled materials.

depth of disassembly. The deeper the disassembly required, the longer it takes and the higher the cost. Ideally, a design should enable quick and thorough disassembly.[18]

1. Avoid permanent connections: Disassembly hinges on connections, which may be reversible or permanent. While reversible connections allow for easy removal and reuse, permanent connections, like welding or bonding, serve to ensure stability. Even so, it is imperative to avoid permanently connecting incompatible materials. In cases in which permanent connections are unavoidable, such as affixing labels to glass bottles, we recommend water-soluble adhesives.
2. Use standardized fasteners: Employing uniform specification bolts in a product streamlines assembly and disassembly processes, enhancing efficiency. Standardizing bolt dimensions reduces the need for changing tools frequently during assembly and disassembly.
3. Reduce fasteners: Unless specific requirements dictate otherwise, minimizing the use of screws or bolts in favor of card connections or mortise and tenon structures simplifies disassembly.
4. Mark separation points and order of disassembly: Employing color, texture, material, or grooves to distinguish separable materials aids disassembly efficiency. Additionally, noting the disassembly order is crucial for taking apart complex components.

Efficient disassembly has emerged as a critical quality and competitive factor in product design. Apple's disassembly robot Daisy exemplifies this, swiftly identifying and disassembling various iPhone models for recycling. Daisy processes components containing aluminum, copper, and rare earth elements, among other materials, with a disassembly speed of 200 units per hour.[19]

4.4 Necessary End-of-Life Processing

Inevitably, all things in the world reach the end of their lifecycles and become redundant. However, what some may consider waste is, in fact, a valuable raw material or an integral component of various ecological processes. Every human action generates reactions that influence the environment in some capacity. Recycling, as a fundamental strategy of sustainable design, aims to repurpose discarded products or materials into raw materials for manufacturing new products. By doing so, it

[18] Disassembly depth is the degree of component separation, which is an important variable in analyzing recycling costs. For details, see: Vezzoli C. Design for environmental sustainability: Life cycle design of products [M]. second ed. Berlin: Springer, 2018:180.

[19] Apple Inc. Environmental Responsibility Report: 2019 Progress Report—A Comprehensive Review of Fiscal Year 2018 [R/OL].[2020-03-15]. https://www.apple.com.cn/environment/pdf/apple_environmental_responsibility_report_2019.pdf

effectively prolongs the lifecycle of materials, contributing to resource conservation and environmental protection.

Nevertheless, it is important to recognize that recycling serves as a remedial measure enacted after items have already been discarded, representing a form of end processing. Furthermore, every phase in the recycling process, from item collection and transportation to cleaning and remanufacturing, incurs environmental impact. Additionally, the quality of materials tends to degrade with successive regeneration cycles. Therefore, while recycling is a key aspect of sustainable design, it is not an ultimate solution. Relying solely on recycling may inadvertently justify excessive consumption and wasteful use of natural resources.

Designers have a responsibility to consistently seek truth and engage in product lifecycle analysis. By tracing materials back to their sources and continuously exploring increasingly effective strategies and methods in design, they can mitigate environmental impacts and strive toward attaining genuine sustainability.

Chapter 5
Design for Reuse and Upcycling

Abstract The design goals of reuse and upcycling are to extend the lifespan of products as much as possible. Reuse involves reusing a product again once or multiple times after its usual period of usage is over. Upcycling, on the other hand, focuses on exploring the inherent characteristics and potential of products and materials to transform discarded items into products of greater value or into entirely new uses.

Keywords Reuse · Upcycling · Item Appreciation · Product-service system

Although reuse and upcycling are distinct concepts, the primary goal of engaging in them is to prolong the lifespan of products. While the recycling measures discussed in Chap. 4 are effective, promoting the reuse and upcycling of products before transporting them to recycling companies is a crucial strategy. By increasing the efficiency, frequency, and length of product use, even in some cases expanding a product's functions, we can maximize its lifespan and increase sustainable value.

5.1 Design for Reuse

Thrift is a natural human trait. Throughout the long and arduous process of evolution, reusing items was a common behavior, while people considered discarding still-useful objects unimaginable and wasteful.

Reuse is straightforward concept: it entails using item once or multiple times after its normal usage period has expired. Reuse involves the use of the item itself (or its parts), not regeneration of its materials. The goal of reuse design is to extend the lifespan of items as much as possible, preventing premature disposal. However, it is not possible to reuse all items, such as certain medical testing supplies. Additionally, we must consider the safety of reuse and overall environmental

X. Liu et al., *Design for Sustainability*, Sustainability and Interdisciplinarity,
https://doi.org/10.1007/978-981-96-8657-5_5

impact. For example, reusing a plastic bottle that leaks toxic substances clearly contradicts the fundamental goals of design.

Both individuals and institutions can practice reuse. In addition to simply reusing objects, individuals can donate or share items they might otherwise discard and also reuse packaging. Companies and institutions can reuse product parts and packaging or urban facilities and buildings. Reuse is not just an ecofriendly behavior, but also a design philosophy. While consumers' initiative and creativity are crucial, professional designers play a vital role in promoting reuse.

Designers can engage in design practices that ease reuse, incorporating interfaces, providing new accessories, and facilitating salvaging items. When doing so, designers must consider the structural strength and durability of products and aim to maximize the level and frequency of reuse, while balancing economic costs and functionality. Examples include designing reusable coffee cups, packaging boxes, multi-purpose plastic bottles, and modular furniture and daily necessities.

Beyond product design, designers can contribute to innovations in circulation services for old items and trading platform systems. They can also create aesthetically pleasing and practical items or decorations from waste (i.e., upcycling), raising public awareness and participation through impactful visuals. However, this latter aspect is supplementary. Compared to non-material design and the creation of more durable, easy-to-dismantle products, upcycling is merely an added benefit.

5.1.1 Daily Item Reuse Design

Many elderly people tend to cherish objects and value frugality, making reuse of things in their everyday lives a natural behavior and habit. Examples include repurposing old cans and water cups, modeling toys from old items, and renovating and reusing old clothes. They may also gift old items to others, or exchange or trade them. Historically, pawnshops not only temporarily stored and traded relatively precious secondhand items, ensuring value retention (though not always at fair prices), but also became part of the financial industry, with trust companies later assuming a similar function.

With economic development and improved living standards, the objects we need for daily life have become increasingly inexpensive and readily available. Advertising has reshaped consumption habits, leading to a rise in disposable items infiltrating daily life. The traditional value of seeing cherishing old things as a virtue has weakened over time. However, the crisis arising from resource shortages and environmental problems caused by overconsumption has raised the alarm, reviving interest in green lifestyles of thrift and frugality. The notion of low-carbon living has gained widespread popularity. Meanwhile, online secondhand trading platforms have flourished, expanding their reach, self-organization, and flexibility.

Despite these positive trends, relying solely on the wisdom of the so-called green experts and spontaneous DIY actions is insufficient to promote the widespread adoption of a make-the- best-use-of-everything ethos in a consumer society. Designers'

Fig. 5.1 Ecofriendly shopping bag launched by British designer Anya Hindmarch (Image source: https://www.anyahindmarch.com)

active intervention and promotion of a change in consumption habits are particularly important. For instance, surveys indicate that every Briton uses an average of 167 plastic bags per year. In response to the environmental impact of all that waste, British designer Anya Hindmarch launched the "I'm not a plastic bag" ecofriendly shopping bag in 2007. Aligning with the lifestyle media and events platform, Green is The New Black, Anya Hindmarch combined environmental protection with fashion, aiming to promote reuse with a multi-purpose canvas shopping bag that reduces dependence on disposable plastic bags. The simple, clear environmental concept in a limited edition available for just £5 became a highly sought-after it bag in the fashion world that year, a classic example of green design (Fig. 5.1).

We use toothbrushes every day. Most people change their toothbrushes every three to four months. With more than seven billion people in the world, you can imagine how many toothbrushes we discard each year. In order to reduce plastic waste, the Touguet toothbrush adopts a modular approach that allows for disassembly. The design is exquisite and simple. Users need to replace only the brush head every few months, while the body of the toothbrush lasts a long time, effectively reducing the production of plastic waste.

5.1.2 Redesign of Disposable Products

Today is an era of disposable products. Think about the disposable items in your life: water bottles, disposable paper cups, takeaway food containers. China throws out 60 million fast food boxes every day.[1] Due to the pandemic, this number has

[1]Wang Qufeng. Comparative Analysis of Kitchen Waste Treatment Technology in China[J]. Guangdong Chemical Industry, 2016, 43(10): 147–149.

undoubtedly surged since 2020. When we stay in hotels, we use disposable toothbrushes, slippers, towels, razors, and so on. Undeniably, disposable products introduce convenience into our lives, but for these conveniences, we constantly deplete precious, limited resources, while polluting the environment.

Some advocate that use of disposable products is rational, particularly in the medical field, citing examples such as syringes, needles, and bandages. The primary arguments for disposability are related to safety concerns and the significant complexity and time costs associated with effective disinfection, making discarding these items after a single use seem justified. However, it is worth noting that, for many years, tools such as infusion bags and needles were successfully reused following proper cleaning and disinfection without significant issues.[2] Therefore, safety is not the core issue but rather considerations of time and money (corporations quietly shift the cost of disposable products onto consumers) and not environmental impact. At least some of the time, we require certain medical and public supplies in disposable form. *The point is that disposable items are increasingly becoming universal in daily life, ubiquitous, upon which we are gradually becoming dependent as a matter of course.* Most people do not feel guilty or remorseful about this behavior. In this collective unconsciousness, we are collectively consuming the planet's energy. How long can we play this game?

In the future, there will inevitably be a trend toward a social consensus regarding the need to rethink and redesign disposable products. When we need disposable items, manufacturing products out of biodegradable materials may be the best answer (a topic discussed in detail in Chap. 7). Current design practice dictates that designers consider disposable products from multiple angles, such as system innovation, structural innovation, and function substitution.

The rise of digital service platforms made takeout popular in China. The resulting use of a large number of disposable food containers has created significant environmental problems. Founded in 2014, Different Good Food was the first restaurant in China to provide plate recycling. Only providing takeaway, the restaurant mainly serves customers within a three-kilometer radius and positions itself as a sort of community kitchen that targets young people who have a fast pace of life and lack cooking time or skills. Different Good Food uses household ceramic tableware as takeaway packaging. Customers do not need to pay a deposit. After the peak dining period, the restaurant reuses the dishes after cleaning and disinfection. As of March 2019, it had five restaurants in Beijing. The original intention of using ceramic plates was to provide consumers a home atmosphere and this became part of the company's core branding. However, it also significantly reduces consumption of plastic takeaway packaging. In early 2019, the restaurant cut plastic packaging by 5000 per day and achieved profitability through strict data management (Fig. 5.2).[3]

[2] Xie Zhuofu. Design Reflection: Sustainable Design Strategy and Practice[M]. Liu Xin, Qin Jingyan, trans. Beijing: Tsinghua University Press, 2011: 183.

[3] March 21, 2019, Beijing Daily http://env.people.com.cn/n1/2019/0321/c1010-30988737.html

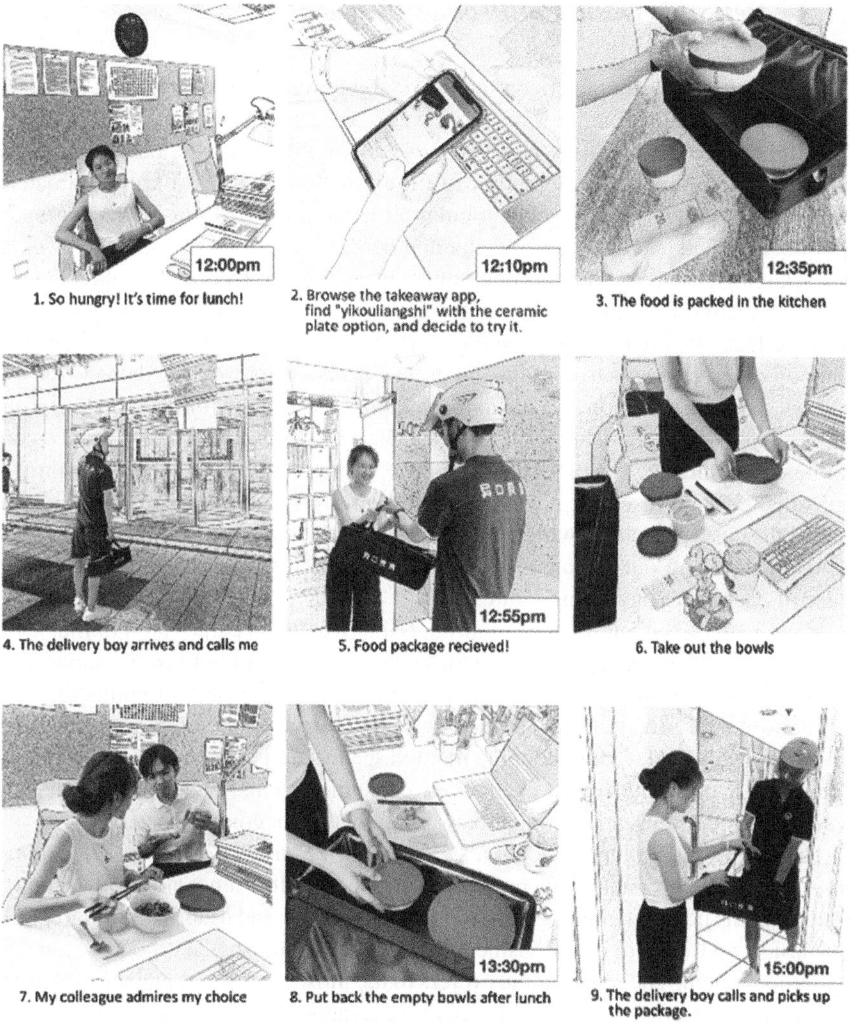

Fig. 5.2 Service process of Yikouliangshi (Image source: Author)

If consumers are environmentally conscious and make changes to their daily habits, then they can bring their own reusable coffee cups. Designers have already launched a variety of products. If managers, merchants, and consumers all make reducing use of disposable items a mission and introduce a variety of alternatives, solving this problem should not be difficult. The massive appearance of disposable items in China is only a matter of the past two or three decades.

5.1.3 Reusable Design of Product Packaging

Unfortunately, most containers on today's market are disposable. Use-and-discard packaging is a major source of resource waste and environmental pollution, especially that made from plastics. Statistics reveal that 95% of plastic packaging loses its economic value after a single use, resulting in an annual loss of approximately $80 to $120 billion.[4] The significant environmental impact of plastic packaging is well-documented and need not be reiterated here.

Packaging functions both as a component of certain products and an independent product. According to the Chinese national standard (GB/T 4122), packaging protects products, facilitates storage and transportation, and promotes sales.[5] The core function of packaging is to contain and protect products, preventing damage during storage and transportation. If packaging remains intact after fulfilling its purpose, reuse seems logical. Reusing packaging can save resources and energy, reduce corporate costs, enhance brand loyalty, optimize operations, and create new commercial value through incentive mechanisms and information technology. However, in reality, the costs associated with recycling, transportation, cleaning, and consumers' habits of throwing out packaging have eroded these potential advantages, making reuse of packaging increasingly rare.

Many Chinese people who lived through the twentieth century are familiar with the practice of "getting soy sauce." In those days, when household consumables like soy sauce, peanut oil, sesame paste, rice vinegar, white wine, and grains ran out, people would take their original bottles, cans, and bags to department stores or grain shops for refills. Today, this shopping method seems cumbersome and inconvenient. This shift is not only due to changes in consumer habits, but also because few designs consider reuse, resulting in packaging that is not conducive to reuse.

The Ellen MacArthur Foundation's 2019 report, *Reuse—Rethinking Packaging*, notes that reusable packaging represents a billion-dollar innovation opportunity. Realizing this potential requires improving its durability and usability through product design and developing a business model that establishes a smooth, valuable operating system. The report mainly refers to the most common container packages, such as those used for laundry detergent, beverages, and condiments. Home and other end scenarios like offices and on-the-go, mid-end scenarios such as retail points are the two core imagined use settings. Reuse entails users refilling containers themselves or returning them to manufacturers for refilling.

Innovative design can promote reuse of other types of packaging as well. Designers can repurpose items using their original materials, structures, and spaces or give them new functions and forms. For example, the Chinese brand Natural Creation designed the Bamboo Wind New Year Tea Gift Box (The World Bohea) in

[4] Twenty-six percent of the total plastic we produce is used for packaging. For detailed data, see the Ellen MacArthur Foundation's 2016 report, "The New Plastics Economy: Rethinking The Future Of Plastics."

[5] Refer to the national standard "Packaging Terminology Part 1: Basics" (GB/T 4122.1—2008)

2023. The woven bamboo package transforms into a beautiful lampshade after the tea leaves are consumed. The packaging may also be creatively repurposed into other aesthetically pleasing, long-lasting household items.

The issue of express delivery packaging presents a significant global challenge. According to a report by Guangming.com, the annual volume of express deliveries in 2020 reached 83 billion pieces.[6] Developing rapidly, the delivery industry exerts tremendous pressure on the environment, especially with express packaging, which has garnered widespread societal attention. In response, the Chinese State Post Bureau formulated the Mail and Express Packaging Management Measures, which emphasize priority use of reusable and easily recyclable packaging.

In 2017, during Double Eleven, the Chinese singles holiday, Suning launched a shared express box project. This packaging utilizes polypropylene (PP) material, known for its lightweight, sturdiness, durability, and non-toxicity, as well as its reusability and recyclability. Calculations indicate that shared express boxes are more economical than corrugated boxes of the same size due to their multiple uses (Fig. 5.3). Concurrently, JD.com, in collaboration with various brands, officially launched the Green Flow Plan, introducing a new type of express box—the Green Flow Box. These boxes are designed to be recycled after delivery and can be reused up to 20 times. If a Green Flow Box is damaged, it can be completely remade. Additionally, the tape used on the Green Flow Box is made from degradable materials, further enhancing its environmental friendliness (Fig. 5.4).

Fig. 5.3 Suning Shared Express Box (Image source: https://www.sohu.com)

[6] See Guangming Daily reporter Dong Bei's report from April 12, 2021, "What to Pack, How to Pack, How to Manage—What Changes Are Coming to Express Packaging."

Fig. 5.4 JD.com Green Flow Box (Image source: https://www.cndsw.com.cn)

In addition to the Shared Express Box and Qingliu Box, other sustainable packaging initiatives like the Box Companion Plan, Cainiao Box, Feng BOX, and Grayscale Box reflect the attention large Chinese e-commerce and logistics companies are devoting to the issue of express packaging, as well as their sense of social responsibility. However, these projects face significant operational challenges. Unlike disposable corrugated boxes, reusable packages require multiple steps, including recycling, sorting, cleaning, disinfection, storage, and redistribution, and demand the active cooperation of couriers and users. Consequently, their overall cost is relatively higher. Packaging is a systemic issue that necessitates not only more durable, sturdy, and convenient packaging box designs, but also improved management mechanisms, service designs, policy promotion, and corporate persistence.

5.1.4 Promoting the Design of Service Platforms for Item Reuse

Although reusing items is crucial, service platforms that facilitate the exchange of old items may have more influence on society and appeal to individual consumers. Since ancient times, places where people can exchange used goods have been integral to daily life. These include pawnshops, secondhand stores, used car markets, and flea markets. Notable examples include Beijing's Panjiayuan Antique Market, Paris's Montreuil Flea Market, Istanbul's Grand Bazaar, and Budapest's Ecseri Market. In addition to these urban markets, there are countless secondhand markets in small towns, which not only provide people with the opportunity to treasure hunt, but also create spaces for interpersonal interactions. These traditional markets often emerge spontaneously and require only moderate management and minimal deliberate design. However, in today's consumer society, opening a secondhand store requires ingenious design and careful creation of the appropriate ambiance.

Today, traditional flea markets still exist, but their scale and popularity are not as extensive as they once were. They have gradually become tourist attractions and nostalgic cultural symbols, while online used trading platforms based on Internet technology are thriving. Information technologies like mobile Internet, Internet of Things, big data, and the establishment of logistics networks provide strong infrastructure support for the rapid sharing, efficient integration, and low-cost trading of resources. Individuals, organizations, and enterprises can post information about things they no longer need through online platforms and even monetize them. The surplus value of idle goods is thus reborn through the redistribution of supply and demand, extending the life of items otherwise likely to be discarded and reducing material resource waste.

The sharing economy has caused the global secondhand trading market to develop rapidly through platforms like eBay in the United States, Mercari in Japan, Finn in Norway, and Xianyu, Zhuanzhuan, Paipai, and Wanwu Xinsheng in China. Their market scale, product and service categories, and user numbers surpass those of traditional flea markets. Several institutions predict that China's secondhand market will exceed 1 trillion yuan in 2021.[7] For instance, Alibaba's Xianyu saw its transaction volume exceed 200 billion yuan in 2020, with 300 million users, a 100% increase from the previous year. The types of no-longer-needed items traded are increasingly diverse, including digital products, clothing, books, and even secondhand luxury goods and large assets like houses and cars. Sales of intangible assets such as knowledge, services, and skills are also increasing.[8] Many platforms focusing on niche fields and themes have formed unique community cultures through the resale of items. For example, the Duozhuayu platform, which focuses on the resale of books, has developed an interactive relationship between booksellers and buyers, established a reading culture, and created a stable consumer market. Over time, it has created unique brand IPs, publications, and peripheral products, and launched recycling services for electronic products and clothing, opening offline experience stores in first-tier cities (Fig. 5.5). The highlights of used item transactions include not only cost-effectiveness, but also social and environmental features that attract the young generation. Generation Z's online consumption habits and green consumption concepts make them the main catalyst in sales of used items.

---

[7] Quoted from TECHWEB's report (April 2, 2021), "Idle Fish: It is estimated that GM will reach 500 billion by the end of this year"

[8] Wang Jiabao, Dun Shuai, Huang Qingyue. When idle resources meet "Internet +" - The popular way of the sharing economy[J]. Enterprise Management, 2016(6): 55–57.

Fig. 5.5 Duozhuayu secondhand online and offline bookstores (Image source: https://www.m.thepaper.cn)

Despite the popularity of online secondhand goods trading platforms, there are concerns. In 2021, many platforms began employing professionals to boost orders and sales. Such commercial operations could undermine the original purpose of selling secondhand goods and potentially encourage blind, impulsive consumption. Sharing used items, originally limited to specific communities, could evolve into cross-regional or even cross-border sales and deliveries, which would increase their carbon footprint and invalidate their environmental benefits. Designers must reflect on and question the dominant logic behind these trends and strive to find more suitable solutions. Leveraging big data, the Internet of Things, and intelligent systems, designers can incorporate various factors including types of used items, their residual value, and ranges of circulation and transactions into business models and service processes. Using tangible resources for local community operations and employing blockchain technology to manage unused virtual resources within communities may be the future of recycling and reuse.[9,10] The goal of reuse is to create a society with no waste and no unused objects, rather than to promote a significant increase in the number of secondhand goods transactions.

5.1.5 *Reuse of Product Components*

Reuse entails companies recycling their products and reusing functional parts, including packaging. This approach not only conserves resources and protects the environment but also is commercially viable. Many companies worldwide have

[9] Zhang Ayuan. The characteristics, dilemmas and explorations of the C2B2C business model in the online used book industry—Taking "Duozhuayu" as an example[J]. Publishing and Distribution Research, 2019(3): 48–53

[10] Song Lifeng, Song Yuanfang, Guo Xiaodan. Sharing of real and virtual idle assets based on data rights—Research on the development of the sharing economy from the perspective of blockchain[J]. The Economist, 2019(8): 39–47.

committed to recycling their products at the end of their life and reusing components that are in good enough condition to manufacture new products. The key to successful reuse is ensuring the safety, functionality, and overall quality (including visual appeal) of the product to allay users' concerns and build lasting brand trust. While consumers may have reservations about products made from used components, increasing public environmental awareness and advances in design and manufacturing are gradually winning consumers over. Support from consumers support will stimulate corporate social responsibility and proactive recycling efforts.

Recycled components often require repair, refurbishment, or remanufacturing before use in new products. Both reusing old components and refurbishing them require early design planning. For example, adhering to principles of *disassembly design* and *modular design* ensures rapid, convenient, safe separation of products and replacement with new components, when needed, facilitating reprocessing and reuse by manufacturers. Additionally, a well-designed recycling system is crucial for encouraging user participation, optimizing processes, and reducing costs. The feasibility of reusing a product or component depends on a comprehensive assessment of factors that include conducting a lifecycle analysis to evaluate the environmental impact and economic costs associated with transportation, cleaning, refurbishment, and remanufacturing.

Flokk, a renowned office furniture manufacturer in Norway, exemplifies this approach with its BMA Axia 2.0 office chair series. Made entirely of recyclable materials with 100% green energy, this ecofriendly product line uses unique assembly technology that eliminates screws, embodying the principle of easy disassembly. Its design likewise facilitates component replacement and maintenance, enhancing reusability. Through the Next Life plan, Flokk recycles old products at the end of their lifespan, maximizing component reuse to manufacture new products and minimizing waste. This series of office chairs won the 2011 Cradle to Cradle (C2C) Award.

Similarly, Caterpillar Remanufacturing Industries (Shanghai) Co., Ltd., a subsidiary of the US Caterpillar corporation, specializes in remanufacturing used construction machinery products.[11] Using advanced engineering technology, it professionally repairs and remanufactures used construction machinery parts, restoring their performance and quality to the level of new products. These remanufactured parts are for sale at a significantly lower price than new parts, helping customers reduce costs, while decreasing the demand for raw materials, conserving resources, and significantly reducing waste and exhaust emissions.

[11] Established in 1925, headquartered in Illinois, in the US, Caterpillar Inc. (CATERPILLAR, CAT) is one of the world's largest manufacturers of construction machinery and mining equipment, gas engines, and industrial gas turbines.

5.1.6 Reuse of Building Materials and Places

Although urban renewal is necessary, it is highly imprudent to demolish old, structurally sound buildings simply to construct fashionable, iconic new ones. This practice leads to the unnecessary waste of materials and energy, alters cityscapes, and results in the permanent loss of cultural and historical memories embodied in old buildings and spaces.

Fortunately, there are intelligent solutions. Rather than demolishing and rebuilding, these solutions respect existing architectural patterns, imbuing them with new functions, meanings, and values through thoughtful transformations. Reuse of buildings has a more significant social impact than reusing industrial. A notable example is the protection and reuse of industrial heritage in Germany's Ruhr Valley.

The Zeitz Museum of Contemporary Art Africa (MOCAA) in Cape Town represents a remarkable example of adaptive reuse. Originally a historic grain silo and once South Africa's tallest building, it was abandoned in 1990. The decision to transform it into an art museum followed considerations of various other uses including as a farm or a parking lot. Heatherwick Studio, the architectural firm in charge of repurposing the structure, executed an extraordinary transformation of the 10-story granary, creating a 4600 cubic meter atrium with stunning spatial form and functional zoning. The deliberately polished arc section contrasts strongly with the rough concrete, creating a dreamlike artistic atmosphere while providing a practical exhibition space.

Chinese architect Wang Shu exemplifies the innovative reuse of materials in architectural practice, integrating cultural context and the emotional resonance of old bricks and tiles into contemporary construction. Since 2000, Wang Shu's team has explored using recycled materials for various projects. One notable practice, derived from traditional construction in Ningbo, employs as many as 80 types of old bricks and tiles to what are called tile walls.[12] For the first construction phase of the Xiangshan Campus at the Chinese Academy of Fine Arts in 2004, Wang Shu's team collected seven million old bricks and tiles from local demolition sites. The Ningbo Museum, completed in 2008, followed the same principle, combining traditional architectural elements with modern forms and techniques, resulting in a design that is both simple and dynamic (Fig. 5.6).

In 2010, Wang Shu used the tile wall method to construct the Ningbo Tengtou Pavilion at the World Expo out of over half a million old bricks and tiles from various villages. This process gave new life to century-old Yuanbao bricks, keel bricks, and ridge tiles. Repurposing waste materials from traditional houses in modern spaces links regions and times, showcasing the practicality, environmental protection, economy, and cultural significance of using waste materials in construction. This practice closely aligns with upcycling. Thoughtful urban renewal that preserves and repurposes existing structures can conserve resources, maintain cultural continuity, and create dynamic, functional spaces. By adopting principles of

[12] Wang Shu. Building Houses[M]. Changsha: Hunan Fine Arts Publishing House, 2016: 46

Fig. 5.6 Ningbo Museum Architecture (Image source: Wang Shu's team)

adaptive reuse and upcycling, architects and planners can significantly contribute to sustainable development.

5.2 Design for Upcycling

German architectural salvage entrepreneur Thornton Kay first introduced the concept of upcycling in a 1994 article in which he interviewed German engineer and reclamation expert Reiner Pilz. When discussing the upcoming EU waste discharge standards, Pilz said, "What we need is upcycling, where old products are given more value, not less."[13] In 1999, Gunter Pauli and Johannes F. Hartkemeyer published a book entitled *Upcycling*, which suggested using nature as a model to ensure non-production of waste throughout the entire economic system and creating value at every stage of material use. William McDonough and Michael Braungart later expanded the concept in their co-authored books *Cradle to Cradle: Remaking the Way We Make Things* and *The Upcycle: Beyond Sustainability—Design for Abundance*, which describe upcycling as transforming existing product design and production methods through recycling or upgrading industrial products and daily necessities.[14]

In 2014, the *Oxford English Dictionary* defined upcycling as "the reuse of discarded objects or materials in such a way as to create a product of higher quality or value than the original."[15] The *Sustainability Dictionary* similarly defines upcycling as "the process of converting an industrial nutrient (material) into something of

[13] Kay T. Salvo In Germany-Reiner Pilz[N]. Salvonews, 1994-10-11(14).

[14] Pauli, G., Hartkemeyer, J. F. Upcycling[M]. Munich: Riemann Verlag, 1999.

[15] Oxford English Dictionaries, https://www.oed.com/?tl=true

similar or greater value."[16] The Ellen MacArthur Foundation categorizes upcycling as a broad recycling strategy, focusing on materials.[17] These definitions highlight that upcycling applies at the levels of both product and industrial material. Scholars argue that "at the product level, object upcycling is one of the most sustainable solutions, positioned between reuse and recycling, as it typically requires little energy input and can eliminate the need for a new product."[18] At the level of industrial materials, McDonough and Braungart advocate for radical design innovation, aiming to achieve the perpetual recycling of materials as proposed in *Cradle to Cradle*.

This section focuses on upcycling at the product level, emphasizing that we should not reduce discarded items to raw materials and subject them to secondary processing methods such as melting and refinement through chemical or physical processes.[19] Instead, we should study the inherent characteristics of various materials and components in terms of their properties, form, and structure to explore their utility, regenerating them in new product forms and transforming them into objects with new functional value. *The core of upcycling is exploring the inherent characteristics and potential of products and materials, transforming items considered worthless into beautiful, valuable objects with new uses.*

In recent years, upcycling has gained in popularity, apparently becoming a new trend in design. The *Cambridge Dictionary* declared it word of the year in 2019.[20] However, there is still a lack of consensus regarding its definition. The theoretical research and design practice around this important strategy in sustainable design are still evolving. Although the following cases are already inspiring, upcycling concept will provide more space for innovative possibilities in future.

In 2019, Chinese designers Zhou Ziren and Su Xinran designed WHECAT, a biodegradable cat nest, as a new upcycling design solution. The cat nest's primary material is straw that otherwise would be burned, which villagers weave and local factories process into the final products. WHECAT caters to the growing demand

---

[16] Available At: http://www.sustainabilitydictionary.com/index.php?s=upcycle [Accessed 16 92,014].

[17] Macarthur E. Towards A Circular Economy: Business Rationale For An Accelerated Transition[J]. Greener Manag International, 2015, 20.

The Ellen MacArthur Foundation's 2015 report, *Towards the Circular Economy: Business Rationale for an Accelerated Transition*, identifies three types of material recycling: functional recycling, which is the process of restoring materials to their original purpose or to other purposes, excluding energy recovery; downcycling, which is the process of transforming materials into ones of lower quality and reduced functionality; upcycling, which is the process of transforming materials into higher quality and increased functionality. See: https://www.ellenmacarthurfoundation.org/towards-a-circular-economy-business-rationale-for-an-accelerated-transition

[18] Sung K, Cooper T, Kettley S. Individual Upcycling Practice: Exploring The Possible Determinants Of Upcycling Based On A Literature Review [C/OL]. Sustainable Innovation 2014, 19TH International Conference, Copenhagen, Denmark, November 3–4, 2014. [2022-03-18]. https://irep.ntu.ac.uk/id/eprint/2559/1/217970-1133.pdf

[19] Circular reconstruction is considered a form of material downcycling. See Chap. 4 for details

[20] See https://dictionaryblog.cambridge.org/2019/11/04/cambridge-dictionarys-word-of-the-year-2019/, the Cambridge online dictionary defines upcycling as "the activity of making new furniture, objects, etc. out of old or used things or waste material"

for pet supplies with a design that fully takes advantage of straw's material characteristics. Traditional artisans weave it into braids through, then coil it into baskets, transforming discarded straw into a biodegradable product that reintegrates into urban life. WHECAT is entirely handmade, so each nest is unique, satisfying the aesthetic tastes of different customers. The product reduces agricultural waste, alleviates air pollution, and increases villagers' employment, providing them additional income (Fig. 5.7).

Laquercia21 is a design studio in central Italy that advocates green design concepts. It uses recycled secondhand furniture, old wooden floors, and even discarded wood from construction sites to design and manufacture furniture. The studio's aesthetics resonate with the timeworn materials, generating a sense of wonder. The designers jokingly say that these pieces of furniture come from grandmother's kitchen and call them naive modernist design. The Credenza Rossa drawer cabinet is typical of their work, with an oak countertop, a red painted body, and recycled

Fig. 5.7 WHECAT Biodegradable Cat House (Image source: www.whecat.com)

colorful old wooden boards as drawer panels. The handles are also items from a secondhand market, creating a unique style. It is more appropriate to call this cabinet a piece of art than a piece of furniture. Each piece of work is unique, with creativity activating the value of old materials.

FREITAG is a leader in upcycling design. Founded by the Freitag brothers in 1993, it is an ecofriendly bag designer and manufacturer based in Zurich, Switzerland. All the raw materials for its bags come from waste, with the main ones being waterproof canvas from redundant shipping container trucks, automobile seatbelts for straps and handles, and bicycle inner tubes. FREITAG's design perfectly interprets the concept of upcycling, turning seemingly ordinary waste materials into fashionable products, greatly enhancing the value of materials, and reducing the energy consumption waste disposal causes. FREITAG also creatively uses leftover materials, designing and producing small accessories such as coin purses and ticket holders from remaining scraps from manufacturing, truly making the most of everything. FREITAG takes full aesthetic advantage of the original paint on the trucks' waterproof canvas, through reasonable cutting, allowing the old rain cloth to shine with fashion, while consciously avoiding pattern repetition in design, ensuring that the surface patterns of the products are almost all unique, satisfying different consumers' aesthetic tastes as much as possible (Fig. 5.8).

The TransGlass Series Ecofriendly Glass Bottle flower vase is a permanent collection in New York's Museum of Modern Art. Since launching, these vases have become Artecnic Design Company's bestselling products. Cutting and splicing

Fig. 5.8 FREITAG Products (Image source: https://www.freitag.ch/zh)

Fig. 5.9 Artecnica's TransGlass Series Vase (Image source: https://tordboontje.com/transglass/)

glass bottles creates vases that are both aesthetic and functional. With recycled materials and smooth, streamlined designs, the TransGlass series are environmentally positive. Combining traditional craftsmanship with artistic design, each TransGlass series ecofriendly glass bottle ornament is a unique work (Fig. 5.9).

In 2019, YUUE Design Studio collaborated with Mobike to implement the Upcycling Shared Bicycles Project. The designers incorporated discarded bicycle parts into their designs to minimize waste and explore the value of the parts' entire lifecycles. The designers cleverly combined various bicycle parts with the result that, after continuous testing, validation, and scheme adjustments, they finally designed a series of home products, including a lounge chair, floor lamp, and coffee table (Fig. 5.10).

Like reuse, upcycling is not exclusive to designers. In 2009, Henrietta Thompson wrote a guide to encourage ordinary people to remake old items and experience a waste-free life. Her book contains sections on furniture, storage tools, lamps/home accessories, textiles/fabric decoration, and cleaning supplies. With simple illustrations and detailed instructions, Thompson teaches readers how to use daily waste to create a home environment that contains a sense of design. This book has a contemporary environmental spirit that is both fashionable and money-saving. In 2012, Thompson released a sequel, *Remake It: Clothes*, introducing readers to designing and fabricating new clothes out of old ones.

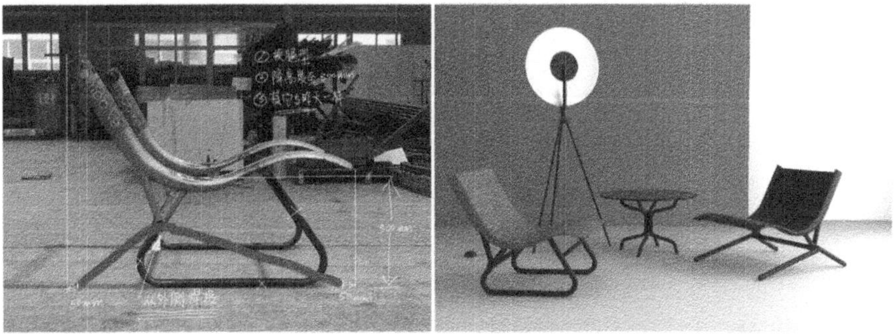

Fig. 5.10 Upcycling Shared Bicycles Project (Image source: https://yuuedesign.com/works/2019-upcycling-shared-bicycle/)

5.3 Item Appreciation and System Remake

As noted above, the concepts of reuse and upcycling overlap significantly during the period when products are in use. Reuse primarily focuses on extending the lifespans of items through improved design, whether by extending their original function or repurposing them as part of changes in user behavior. In contrast, upcycling builds upon reuse by enhancing the value of discarded materials through creative design interventions. This approach examines materials, functions, and forms to explore the full utility and potential worth of waste items.

For example, transforming a glass wine bottle into a stylish vase through cutting, grinding, and reassembling exemplifies upcycling, as it elevates the item's value beyond its initial purpose. Some designs, like Thomas Leech's Shoey Shoes, combine multiple strategies by transforming waste from the leather industry into children's footwear, thereby significantly increasing material value. This design also incorporates a modular concept to facilitate reuse of parts and simplify recycling.[21]

From a theoretical standpoint, upcycling serves as a corrective strategy to prevent downcycling, thus intervening post-production to enhance material value.[22] This strategy relies heavily on designers' sensitivity to waste and their ability to develop personal, artistic concepts. Unlike approaches suitable for mass production, upcycling thrives on conceptual flexibility. Because upcycling cannot occur on a mass scale its implications for sustainability require careful evaluation. However, achieving an upgrade in the redesign of individual items is insufficient, but instead requires a system remake.

In *Cradle to Cradle*, William McDonough and Michael Braungart first proposed the concept and practice of recycling. After more than 10 years of research and

[21] Reference: www.thomasleech.co.uk/#/shoey-shoes

[22] Once the problem and harm are realized, mitigation and remedial measures only reduce the intensity of harm to a certain extent and prolong the cycle of harm. See Chap. 2 "Sustainable Design"

exploration, they published *The Upcycle: Beyond Sustainability—Design for Abundance*. There, they state that we should no longer focus on how to reduce human impact and harm to the ecosystem, but fundamentally redesign human activities.[23] Their concepts of upgrade and reconstruction transcend conventional definitions. They advocate for redesigning human activities to establish a virtuous cycle within ecosystems, moving beyond mere reduction of environmental harm toward achieving abundance for both humans and nature. This perspective challenges traditional environmentalism by proposing sustainable design solutions that enable lifestyles compatible with ecological principles without sacrificing personal freedoms or economic growth.[24] For instance, enhancing water sustainability through technologies like filtration and solar heating allows individuals to maintain indulgent habits like taking long, hot showers guilt free, demonstrating that ecological technologies can support more sustainable lifestyles without compromising personal comfort or environmental responsibility.

The discussion touches upon broader societal implications that exceed the scope of the chapter, but invite readers to contemplate lifestyle choices and their intersection with ecological technology. This prompts critical reflection as to whether sustainable development can accommodate diverse lifestyles and ensure equitable opportunities for all individuals.

[23] Mcdonough W, Braungart M. The Upcycle: Beyond Sustainability-Designing For Abundance[M]. Berkeley, CA: North Point Press, 2013

[24] Wang Hongwei, Lin Bo. Ecological Design—Review of "The Upcycle: Beyond Sustainability—Designing for Abundance"[J]. *Design Art Research*, 2015, 5(5): 39–42, 48.

Chapter 6
Emotional Durability Design

Abstract Emotional durability design entails creating products that are more robust, less prone to damage, and less likely to become outdated during their normal use period than conventional ones. Alternatively, such designs can create emotional connections that make consumers reluctant to discard them. This approach offers a strong critique of and correction to fast fashion and consumerist culture.

Keywords Emotional durability · Planned obsolescence · Emotional design

Similar to the goals of reuse and upcycling discussed in the previous chapter, the objective of emotionally durable design is to extend the useful life of items as much as possible, thereby avoiding premature disposal. However, emotionally durable design is a proactive strategy that focuses on developing products that are inherently robust and not prone to damage during normal use or that generate strong emotional connections and therefore become difficult to discard. In contrast, reuse and upcycling emphasize extending the utility of items beyond their normal use period, either by continuing their original function or by creating new uses and values.

6.1 Reasons for Discarding Items

We discard things for various reasons. Often, we dispose of still-intact, useful items such as furniture, appliances, decorations, and, notably, clothing. While decluttering may serve as a justification, in practice, we frequently forget this impetus and acquire new items. Therefore, before discussing durable design, we need to understand why items are not durable. Generally, there are several reasons for abandoning everyday items:

1. *Product Aging:* Normal use leads to performance, structural, or material aging. All products have a lifespan. Abandoning a product that is no longer safe or reliable is often justified. However, design choices determine a product's lifespan.

X. Liu et al., *Design for Sustainability*, Sustainability and Interdisciplinarity,
https://doi.org/10.1007/978-981-96-8657-5_6

Can we select materials and technologies suitable for designing products with increased use periods and durability?

2. *Accidental Damage*: Damage caused by improper use or accidental factors. If a robust repair service system were in place, we could effectively avoid abandonment due to accidental damage. However, repair and maintenance services are often not profitable. In a fast fashion, low-price, competitive market, except for certain high-end products, there is little incentive for merchants to provide repair services. How can we reduce such damage through design? Can we design products that are easy to repair and maintain? Or can we develop effective service systems? Addressing these questions will help achieve durable design.
3. *Technological Obsolescence*: Technological advancements render original functions obsolete. Most information products are subject to Moore's Law and require continual updates and replacements.[1] For example, despite the sturdiness of old phones, the allure of new features, improved cameras, and increased battery life make new phones irresistible.
4. *Psychological Factors*: Product abandonment occurs due to aesthetic and cultural factors, also known as psychological obsolescence. Fashion trends and advertising can make users feel their products are outdated in terms of style, color, material, or decorative patterns. This kind of abandonment is deliberately designed and serves as a fundamental factor affecting product durability.

Most items are discarded not due to technical failures or aging. In the UK, 33% of household appliances are thrown out while still functional.[2] While data on China is lacking, the impact of consumerism suggests a similar trend. Globally, consumers discard wearable clothing worth up to $460 billion annually, with some items worn only 7–10 times.[3] Alarmingly, significant quantities of clothing are never sold and simply destroyed. It is distressing that people wear only for a short time or not at all clothing manufactured at a huge environmental cost.[4]

[1]Gordon Moore, one of the founders of Intel, proposed Moore's Law. It states: when a price remains unchanged, the number of components an integrated circuit can accommodate will double every 18–24 months and performance will also double. In other words, computer performance that can be bought for one dollar will double every 18–24 months. This reveals the speed of progress in information technology. (Source: Wikipedia https://zh.wikipedia.org)

[2]Cooper T. Inadequate life? Evidence of consumer attitudes to product obsolescence [J]. *Journal Of Consumer Policy*, 2004, 27: 421–449.

[3]Ellen MacArthur Foundation. New Textile Economy: Reshaping the Future of Fashion [R/OL]. 2017[2021-09-26]. https://archive. ellenmacarthurfoundation.org/assets/downloads/New-Textile-Economy-Reshaping-the-Future-of-Fashion _180926_152316.PDF.

[4]In 2018, Burberry burned about 250 million yuan worth of new clothes. Since 2013, H&M burned 60 tons of unsold clothing in Denmark, and in 2017, it destroyed more than 15 tons of inventory. The reasons for this destruction merit deep reflection about the business model today's fashion industry uses. See: Baraniuk C. Will Fashion Firms Stop Burning Clothes? [EB/OL]. 2018-09-17[2022-03-10]. https://www.bbcearth.com/news/will-fashion-firms-stop-burning-clothes

6.2 Planned Obsolescence and Eternal Design

Planned obsolescence is a marketing term that denotes manipulating the lifespan of a product to render it obsolete quickly, shaping a new consumer culture of new vs. old to boost sales. In *Design is the Problem* Nathan Shedroff writes: "The term planned obsolescence is the second most dangerous concept ever invented by marketers. It encourages us to give up or throw away perfectly good things in favor of others simply because we're led to believe that they're no longer useful or fashionable. The phrase was coined in the late 1920s, but it was popularized by an industrial designer."[5] In 1954, American pioneer industrial designer Clifford Brooks Stevens took the concept of planned obsolescence to the extreme, propagating it widely, with the idea that design should instill in buyers a desire to own something a little newer, better, and faster than they actually need. The term became popular beginning in the 1950s as a winning strategy in business marketing. Especially in the United States, this design concept quickly influenced almost all product areas, including cars, home appliances, and clothing. In 2005, Glenn Adamson published *Industrial Strength Design: How Brooks Stevens Shaped Your World*, which comprehensively describes Brooks Stevens' approach and the astonishing number of commercial designs he created.[6]

The core of planned obsolescence is creating psychological obsolescence[7] by making consumers feel insecure about their outdated products, driving them to follow fashion trends and purchase new items. Dr. Jonathan Chapman argues that consumerism is the root cause of the sustainability dilemma, rendering scientific analyses and sustainable design measures ineffective.[8]

Planned obsolescence has had a huge impact on consumers worldwide and generated enormous corporate wealth. It is an unsustainable model for business and consumption that deviates from the original intention of design—to provide creative solutions to actual problems and meet real needs. It remains hugely influential, and its harm to the environment has increased over the past 70 years.

---

[5] Shedroff. N. Design is the problem: The future of design must be sustainable[M]. Translated by Liu Xin, Qin Jingyan. Beijing: Tsinghua University Press, 2011: 168.

[6] Adamson G. Industrial Strength Design: How Brooks Stevens Shaped Your World[M]. Cambridge, MA: MIT Press, 2005.

Cooper T. Inadequate Life Evidence Of Consumer Attitudes To Product Obsolescence.[J] *Journal Of Consumer Policy*, 2004, 27(4), 421–449.

[7] Chapman. Emotional Sustainability Design: Products, Experiences and Empathy[M]. Translated by Lu Mingming. Nanjing: Southeast University Press, 2014.

Martins J, Simões J, Franqueira T. Sustainable Design: The Durability Of Design Classics As A Stimulus To Reduce The Environmental Impact Of Products[C/OL]0.2015[2022-03-10]. https://www.plateconference.org/sustainable-design-durability-design-classics-stimulus-reduce-environmental-impact-products/

[8] Cases I Sampere N. Making More Durable And Reparable Products: Building A Rating System To Inform Consumers And Trigger Business Innovation [R/OL]. 2015[2022-03-10]. http:// makeresourcescount.eu/wp-content/uploads/2015/07/durability_and_reparability-report_final.pdf

A study by Portuguese professors of 100 products that comply with the concept of eternal design and are still in production reveals they share common traits: They have almost no decoration, are simple, easy to use, elegant, lightweight, well-crafted, solid, balanced, human-friendly, comfortable, relatively inexpensive, award-winning, popular, flexible, and minimalist.[9] These products, often traditional wooden, metal, or glass items, exemplify the timeless appeal of thoughtful design.

Prime examples include the Semi-Pendent chandelier Danish design master Claus Bonderup designed for Thorup in 1968, or the simple, comfortable, and sturdy Alfi chair Jasper Morrison designed in 2015 using beech, polypropylene, and wood fiber composite, which may be used for several generations.

Although the study was limited, its conclusions are nonetheless inspiring. While the world today is on the high-speed train of electronic technology and information networks, changing rapidly, classic products still convey eternal charm. As designers attempting to create an improved life, we should incorporate this eternal elegance into future product design.

Time will tell whether the concise forms can achieve eternal aesthetic appeal. Beginning in 1926, Danish designer Poul Henningsen began designing the PH series of lamps. The metal plates on the same axis but at different angles shield the lamp's glare, making the light uniform and soft. Its design is practical, elegant, and exquisite, combining functionality and aesthetic value. The PH lamp is one of a few classic designs still in production today. See Fig. 6.1.

6.3 Durability Design

Durability is the cornerstone of product lifecycle extension. In its 2015 report, the European Environment Agency defined product durability as the maximum potential use life of a product until it no longer maintains its primary function, lacks economic feasibility for repair or part replacement, and is no longer a candidate for adjustment, personalization, or upgrading.[10] Simply put, durable goods are products with an extended use cycle that retain substantial value even after prolonged periods.

The environmental benefits of durable products are significant. They mitigate waste by preventing premature disposal and thus reduce resource consumption and the carbon emissions associated with the production, storage, transportation, and sales of replacement products.[11] The following specific strategies can ensure better

---

[9] Vezzoli C. Design For Environmental Sustainability: Life Cycle Design Of Products[M]. 2nd ed. Berlin: Springer, 2018: 124.

[10] Schifferstein H. N. J., Zwartkruis-Pelgrim E. P. H. Consumer-Product Attachment: Measurement And Design Implications[J]. *International Journal Of Design*, 2008, 2(3):1–13.

[11] Mugge R. Product Attachment[D]. Delft, Netherlands: Delft University of Technology, 2007.

Ceschin F, Gaziulusoy I. Design For Sustainability: A Multi-Level Framework From Products To Socio-Technical Systems [M]. London: Routledge, 2020.

Fig. 6.1 One of the PH series lamps designed by Poul Henningsen (Image source: https://interioricons.com)

achievement of the goal of durable design. They include product reliability, adaptability, and optimized service.

6.4 Reliability

Durability is the most direct proof of product reliability. It requires a judicious selection of materials, optimization of structural strength, and treatments to prevent surface corrosion. Simplifying a product's design and reducing its number of parts also enhances reliability. Consistency in technical performance is likewise critical to durability. It necessitates sound internal system design and manufacturing and assembly of electronic components.

The world's longest-lasting light bulb, the Centennial Light (Fig. 6.2), located at the Livermore Fire Department in California, USA, has been lit since 1901 (with brief outages due to power failures). However, a product's extended lifespan can sometimes conflict with economic interests. In 1924, a consortium of major bulb manufacturers decided to limit bulb lifespans to under 1000 h to stimulate sales.

Fig. 6.2 The Centennial Light (Image source: https://www.centennialbulb.org)

In some cases, durability is not a technical issue. For example, cargo bicycles remain a crucial means of transportation in Africa due to their simplicity and durability. Because they are not high-tech products and have slim profits, Western companies have basically withdrawn from this field. Beginning in the 1970s, African countries started importing Chinese bicycles, valued for their affordability, durability, and ease of maintenance. Historically, the reliability of Chinese manufacturing suffered due to cost-cutting measures, leading to frequent replacements and significant resource waste. However, recent improvements in the quality and reliability of Chinese products have reversed this trend (Fig. 6.3).

6.5 Adaptability

Regardless of a product's reliability, normal wear and tear or accidents will eventually occur. Moreover, technological advancements, societal changes, product repositioning, human physical conditions, and cultural shifts eventually render products obsolete. Adaptability in design can prolong products' lifespan by making them easy to disassemble, assemble, adjust, repair, and upgrade.

Modularization enhances efficiency by allowing independent assembly of different functional modules and standardized parts, thus facilitating fast, efficient maintenance and reducing labor costs. Modular upgrades are particularly important for frequently updated electronic products, enabling the replacement of advanced parts while preserving the majority of a product's components. Such upgrades reduce

Fig. 6.3 Chinese-made bicycles used to transport agricultural produce in Uganda (Image source: https://www.istockphoto.com/hk, downloaded on 20th, Aug, 2023)

environmental impact, allow for adaptations to changing needs, and minimize resource consumption.

Modular design also supports product longevity. For instance, Herman Miller's Mirra office chair offers a 12-year warranty and features easy disassembly for part replacement and repair, ensuring quality and durability (Fig. 6.4). Similarly, shirts from the 1960s and 1970s with replaceable collars and sleeves exemplify modular design, extending clothing life and yielding savings for consumers. Although this practice waned, recent trends in sustainable fashion have included interest in modular clothing, presenting new opportunities for developing sustainable design.

Dynamic size is a sensible approach to sustainable design. Although its range of adjustment is limited, this approach can significantly expand the interval during which a product is viable and extend its service life. Typical examples include teenagers' apparel. Mothers report that their children grow rapidly during primary and secondary school, particularly in terms of foot size, leading to the need to purchase them new shoes often. This makes designing a growing shoe a prudent idea. The GroFive Footwear team in the US designed Expandals, which can grow five sizes and accommodate children aged two to 15. Similarly, the Stokke Steps child seat,

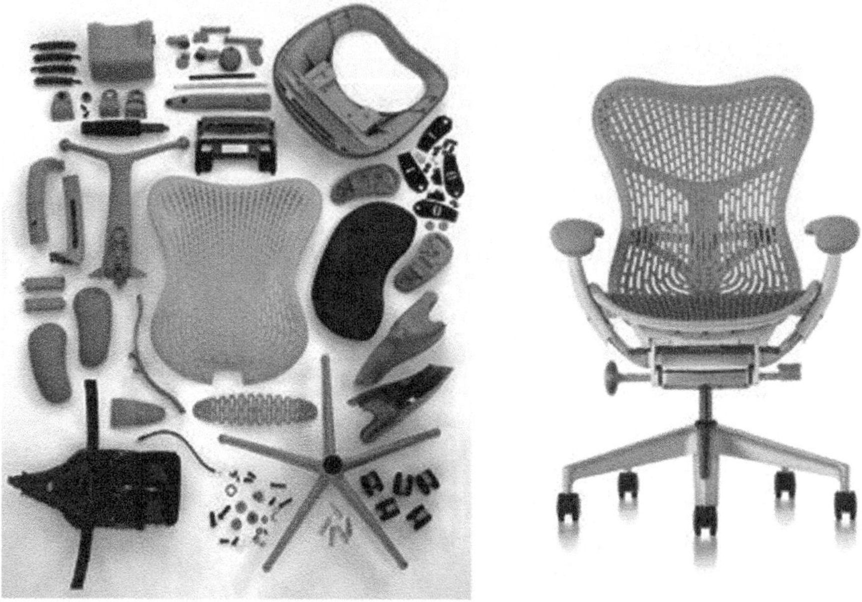

Fig. 6.4 Herman Miller's Mirra Office Chair (Picture source: https://www.hermanmiller.com)

Fig. 6.5 Petit Pli Children's Clothing

created by Permafrost design studio, can adjust to children of various ages, prolonging its utility. Figure 6.5 depicts Petit Pli children's clothing with variable sizes, designed by Ryan Mario Yasin.

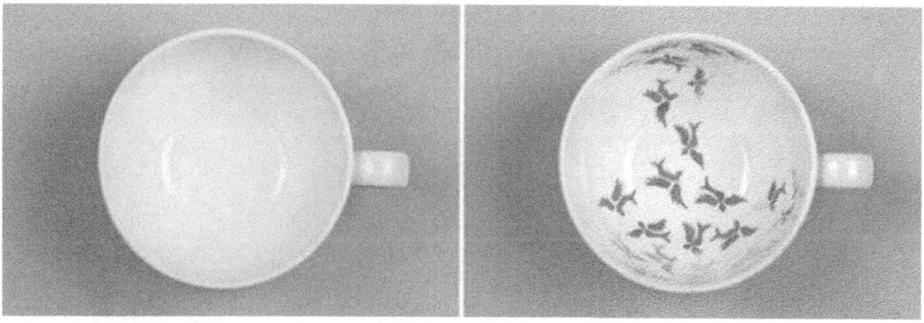

Fig. 6.6 Stain's emotional design cup (Image source: http://www.bethanlaurawood.com)

Multifunctional design seeks to provide as many function options as possible to meet the different needs of various users. For example, a Swiss Army knife, in addition to containing one or more knife blades, includes a bottle opener, nail clippers, screwdriver, and other tools

1. *Design Classic Products with Traces of Time:* Choosing materials or surface treatments that highlight the traces of time and use can strengthen the sense of history embedded in an object and create an anti-trendy, eternal style. For example, Stain's design cleverly turns use into a process of new revelations. Over time, the cup acquires a new aesthetic, continuously stimulating the user's interest (Fig. 6.6). The longer you use it, the clearer the pattern inside the cup becomes. This design by Stain establishes an emotional connection. Even when the cup is old, users will not want to throw it away and may even consider the old cup better than a new one.

Japanese designer Naoto Fukasawa designed the SIWA series of bags using traditional washi paper. After processing, this paper becomes durable, water and oil resistant, and has good load-bearing capacity. The most significant feature is that the paper can record the user's usage story. Over time, unique, complex, aesthetically pleasing textures appear on the fabric. The user's interaction with the bag is akin to painting on it, resulting in a design that "becomes more beautiful the more it is used."

2. *Design Products that Contain Memories and Generate Stories:* As users interact with a product over time, new elements or narratives can arise that embed personalized experiences and memories in the product. Robot pets are an example of this type of product design that provides users with a nurturing experience similar to that found in organic interactions. In 2019, the Japanese company Groove X launched Lovot, a family-friendly robot pet designed for emotional communication and therapeutic effects. Lovot is not only aesthetically pleasing but also highly intelligent. Equipped with 50 sensors and deep learning technology, Lovot can recognize its owner, communicate, exhibit jealousy and attention-

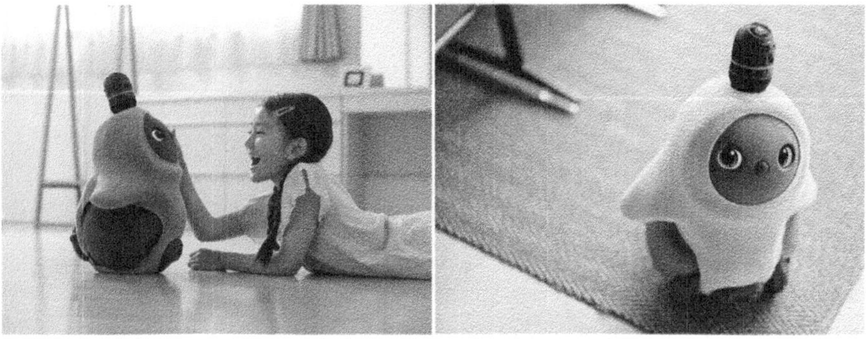

Fig. 6.7 Robot pet Lovot by Japanese company Groove X (Image source: https://store.lovot.life)

seeking behavior, and respond appropriately to environmental stimuli. Weighing precisely three kilograms, Lovot is around the weight of an average baby, making it suitable for picking up and hugging (Fig. 6.7). Lovot is not unique. Other products like Sony's Aibo robot dog, Toyota's Kirobo, and Honda's Asimo are also on the market. In the future, AI-based intelligent robot pets will become integral companions, recording and learning human habits and emotions, and gradually generating new memories and stories through interactions. With software upgrades, they will become increasingly intelligent, while their hardware will remain a family member, to which people will grow close over time and thus be unlikely to discard.

3. *Design Products that Adapt to Changing Needs:* Encourage users to modify products over time to meet their evolving needs. This approach transforms users into co-creators of their experiences instead of merely passive observers.[12] Modular design and assemble-it-yourself furniture exemplify this strategy. The patent for the first generation of LEGO bricks, the classic, purest, and most versatile modular toy, was granted in 1958. Contemporary LEGO bricks are still compatible with those early models. LEGOs are highly adaptable, serving multiple purposes beyond that of mere playthings. From them children can construct cartoon animals and other characters, Gundam warriors, the Tower of London, skyscrapers, or Star Wars spacecraft. Additionally, LEGO bricks function as educational tools for learning mechanical principles, building transmission models, and even combining with electric motors and sensors to create moving vehicles and robots. They can also serve as everyday objects, such as lamps, pen holders, coasters, or decorations, and can even be con-

[12] Chapman. Emotionally Durable Design: Products, Experiences and Empathy [M]. Translated by Lu Mingming. Nanjing: Southeast University Press, 2014.

structed into play structures like caves and castles. LEGO's ability to ignite childlike imagination, creativity, and joy helps establish a profound emotional connection with users.

4. *User Participation in Product Design, Manufacturing, and Maintenance:* Encouraging users to engage in designing products can significantly enhance product personalization and future product satisfaction. When users contribute to assembly and manufacturing, even partially, they infuse products with their emotions, thus increasing their attachment and goodwill toward the resulting objects. Additionally, allowing users to participate in repairs can extend the functional lifespan of products and promote emotional attachment through product personalization. This strategy also provides users with a sense of satisfaction and personal achievement. IKEA, in collaboration with UNICEF, holds the Soft Toy Drawing Competition annually to support children's education projects. Children worldwide can submit their designs, with six winning entries selected for production as part of the SAGOSKATT series of soft toys. These toys, designed by children, are well-received on the market, and the profits from their sales benefit children (Figs. 6.8 and 6.9).

Despite the existence of various technical approaches to apply these strategies, their implementation requires designers to acquire a deep understanding and appreciation of human nature and to project their genuine emotions into product design. In fact, emotional design begins with understanding the psychology of consumers, striving to build strong human–object affective relationships and resisting the indifference and fickleness consumerism imposes.

Fig. 6.8 Camel, candy, brown bear, earthman, eggplant car, and ladybug—all designed by children entering IKEA's annual Soft Toy Drawing Competition (Image source: https://www.ikea.com)

Fig. 6.9 "I drew a camel, my favorite animal. I tell my stuffed toy to always be with you, to keep you company. Give a lot of love to the camel because it's special."—Christel (8 years old) (Picture source: https://www.ikea.com)

6.6 Reflections on Durability

Emotion-based durable design does not always succeed. Emotional attachments between human and objects are highly subjective, making it challenging to precisely customize designs to cater to broad consumer needs. What may be an attractive product to some individuals, others might consider mediocre. Furthermore, a person's tastes or self-image can change over time due to environmental influences and fashion trends in ways that make it challenging for designers to accommodate. Moreover, for some highly functional products (such as washing machines), the impact of emotional design on their service life is minimal.

Additionally, not all products improve with prolonged use and durability is not a universal good. Like any other design strategy, moderation is key. For electronic products that experience rapid technological developments, durable design may not make sense. Beyond this, although modular design facilitates component updates or upgrades, technological change is often revolutionary and systemic. For example, early laptops and mobile storage devices cannot keep up with new system environments and inevitably phase out. For products that have significant environmental impacts during use, such as automobiles, timely updates and replacements are necessary. A car's typical service life is about 10 years, during which technology advances rapidly, significantly improving the energy-saving and environmental protection performance of new models. Extending the service life of old models will

Fig. 6.10 Grandma Catherine and her best friend, Beetle Annie (Image source: http://newsroom.vw.com)

exacerbate environmental damage, if new ones have improved environmental and energy performance. Updating internal machinery and equipment while retaining a classic appearance, exemplified by Grandma Catherine, who has kept her Beetle for 52 years (Fig. 6.10), is a meaningful attempt at finding a balance.[13]

Furthermore, for temporary and disposable items (such as newspapers, wrapping paper, or disposable dishes), extending their service life is less beneficial than applying other sustainable design strategies, such as digital publishing, reusable packaging materials, and biodegradable food containers.

As a sustainable design strategy, durable design relies not just on designers' aspirations, but also on corporations' recognition and active participation. Most companies have little incentive to adopt durable design strategies because doing so would reduce the frequency at which consumers replace products, impacting sales and profits. However, durable products can achieve market advantages and enhance word-of-mouth, brand value. With appropriate product service systems, durable design can encourage enterprises to transition from merely selling products to providing comprehensive solutions, thus acquiring new competitive advantages and sustainable development potential.

[13]The story of "Beetle Annie": Grandma Catherine bought a 1966 Volkswagen Beetle when she was 21 and named it "Annie." The car has driven 560,000 kilometers. Although it still runs, it is now in its twilight years. The Volkswagen North America team spent 11 months carefully restoring it, replacing 40% of the parts. They meticulously refurbished its interior and exterior and made a series of modern upgrades, giving Annie a new look. Source: Volkswagen official website: http://newsroom.vw.com

Chapter 7
Design for Reduction

Abstract Reduction is the most critical of sustainable design strategies and the closest to source control. If we ignore reduction at the source, other strategies such as durability design, reuse, upgrade remanufacturing, and circular remanufacturing will not be fully effective.

Keywords Reduction · Material consumption · Energy consumption · Harmless design

"Design for Reduction" means reducing the consumption of materials and energy throughout the entire life cycle of products, services, and activities, while also minimizing the emission of harmful substances. In other words, design for reduction involves decreasing material and energy use during input: in products, services, and activities throughout their lifecycle, and during output: reducing emissions of waste and harmful substances. Design for Reduction also has social and ethical significance because it forces us to reconsider consumerism and attend to the spiritual evolution of our lifestyles. Owning many pairs of fashionable shoes may no longer signify material wealth but instead spiritual poverty. While technology will eventually be omnipotent, enjoying its fruits and material abundance requires caution and moderation, which may be crucial for sustainable survival in future.[1]

Reduction does not necessitate a poor quality of life. Less is more—reducing unnecessary consumption of materials, energy, manpower, and funds while decreasing pollution of the environment can yield new prosperity, benefits, and welfare. For countries, enterprises, and individuals, it offers a long-term, valuable, meaningful return.

[1]Yuval notes in *Sapiens: A Brief History of Humankind* that "the deterioration of the ecological environment does not mean resource shortage… In our future, we may see Homo sapiens sitting on various new raw materials and new energy, but at the same time destroying the remaining natural habitats, leading most other species to extinction." Yuval, Noah Harari. *A Brief History of Humankind: From Animal to God* [M]. Lin Junhong, trans. Beijing: CITIC Publishing House, 2017: 344.

X. Liu et al., *Design for Sustainability*, Sustainability and Interdisciplinarity, https://doi.org/10.1007/978-981-96-8657-5_7

From a designer's perspective, reduction means developing suitable and durable products that fully satisfy users' physiological and psychological needs, thus reducing frequent replacements and abandonment of unsuitable, unsatisfactory products. This aligns with the durability and emotional design principles mentioned in Chap. 6. Reduction involves a discussion of social, economic, philosophical, and environmental values. This chapter focuses on specific design strategies to reduce environmental impact, particularly reducing consumption of materials and energy, as well as the amount of pollution emitted during production and consumption, and proposes inspiring design solutions.

7.1 Reducing Material Consumption

Reducing material consumption involves minimizing the use of materials at all stages of a product's lifecycle. This strategy has significant environmental benefits, as human production activities involve input of material resources and output of waste. Reducing resource input is beneficial to protecting the planet's increasingly depleted natural resources. Reducing the volume of materials a product requires has economic benefits because it lowers materials and processing costs, as well as transportation load.

Designing to reduce material consumption often starts with product miniaturization and lightweighting, and progresses toward dematerialization,[2] such as using digital technology or service design methods to replace physical products. This transformation is gradual and cannot be achieved overnight.

Technological progress is crucial for reducing material consumption, especially because the rapid development of electronic and network technologies makes internal modules more compact. Network cloud platforms can eliminate the need for internal storage, contributing to the continuous slimming of products. However, miniaturization and weight reduction are not goals in themselves. Usability remains paramount. Reductions in product size must adhere to basic ergonomics requirements and people's habits of use. For example, from a technical point of view, the size of a mini USB flash drive can be astonishingly minute. But if you think about your own fingers and the way you plug and unplug a flash drive, you understand that small is not the goal. Suitability is the standard of good design.

The following sections discuss strategies for reducing material consumption in product manufacturing, packaging, and use.

[2] Design for Dematerialization refers to reducing the overall size, weight, and material used in designs to reduce environmental impact. This is also an important strategy in circular economy design.

7.1.1 Product Manufacturing

Designing products with weight reduction in mind significantly diminishes the environmental impact a product has during manufacturing. Most products have potential for weight reduction, but a lack of awareness, motivation, knowledge of new materials and processes, or concerns about risks hinder downsizing. In addition to clever design ideas and collaboration between designers and engineers, encouraging policies, pressures from competitors, and an internal drive for excellence are crucial for creating change.

MIO Design Studio's Bendant Lamp embodies the weight reduction principle. Made of recycled metal and utilizing laser cutting technology to minimize waste, its two-dimensional, flat form saves space during transportation and packaging. Consumers can create their own product shapes and light effects through simple folding (Fig. 7.1).

1. *Develop small, light products:* Small, light products save on materials and reduce the cost of logistics. Their decreased size and weight allow for transportation of an increased number of products using the same number of freight handlers, reducing overall carbon dioxide emissions, and saving fuel.
2. *Reduce the number of parts and types of materials:* Simplifying products saves resources and streamlines the complexity of manufacturing, likewise easing assembly and disassembly. It also facilitates material recycling of materials, as Herman Miller's Celle office chair demonstrates.
3. *Reduce complexity:* Avoid processes like electroplating, printing, hot stamping, and in-mold printing, which increase the difficulty of manufacturing and recycling, costs, and environmental impact. Processes like electroplating can lead to leakage of toxic substances such as heavy metals into the environment. Environmentally conscious design can eliminate or reduce such processes, simultaneously promoting new minimalist trends.
4. *Optimize structural design:* Smart structural design, like reinforcement ribs and bending structures, can increase the strength of a single material, thereby reducing the number of types of materials a product contains, saving on materials and reducing product weight.

Fig. 7.1 Bendant Lamp by MIO Studio (Image source: https://www.mioculture.com)

5. *Use simulation systems and simulation technology:* Simulating design and manufacturing processes helps identify and optimize issues, reducing material waste and time consumption. This technology is increasingly widespread.
6. *Choose reasonable production processes:* Efficient processes improve production and reduce waste and offcuts. For example, 3D printing technology can significantly cut material consumption during processing, while reducing the number of parts and types of materials.

7.1.2 Product Packaging

Packaging's primary function is to protect products during transportation and storage, thus extending their lifecycle. Packaging is also an important means of conveying corporate brand imaging and product information. However, packaging often becomes too important as a marketing tool, leading to excessive use of materials, oversized packaging, and luxurious decoration, like the mooncake gift boxes on the market that are over-packaged under the banner of cultural creativity. These far exceed the basic needs of product protection and information transmission, becoming commercial gimmicks that stimulate consumer desires and cater to vulgar tastes. Therefore, good packaging design is not only a technical improvement, but also a cultural one.

1. *Design simple, lightweight packaging:* Minimize material use and packaging volume while ensuring functionality. IKEA's flat packaging and Apple's minimalist iPhone packaging are good examples.
2. *Design reusable packaging:* This can include creating refillable laundry detergent bottles and reusable courier boxes or beer crates, as well as repurposing packaging, such as transforming used bottles transformed into vases (Fig. 5.9) or teacups.
3. *Use packaging as part of the product:* Packaging can serve as part of the product, such as traditional ice cream cones or natural grain coffee cups.
4. *Design combinable, stackable packaging:* Modular or standardized packaging saves storage and transportation space. Unique shapes for novelty can consume more materials and space, reducing efficiency.
5. *Use recycled materials for packaging:* Most packaging has a single function and limited material quality requirements (except for food-grade packaging), making recycled material components suitable. Recycling packaging materials multiple times using an effective recycling platform is optimal (see Chap. 4).
6. *Design a mechanism to reduce or even eliminate packaging:* Selling products in bulk or concentrated form encourage consumers to reuse packaging. For instance, the German Original *Unverpackt* (unpackaged) supermarket, which opened in 2014, sells locally produced organic food, daily necessities, and cosmetics in bulk (Fig. 7.2) without packaging, guiding customers to bring their own containers.

Fig. 7.2 German Bulk Supermarket *Unverpackt* (Image source: https://original-unverpackt.de)

7.1.3 Product Use

Because different products vary significantly in terms of their material consumption during use, we cannot generalize regarding this matter. As mentioned above, furniture products consume the most materials during the manufacturing stage, with negligible consumption during use. Conversely, other products, such as flush toilets (water), vacuum cleaners (filters), and printers (ink or toner), require substantial quantities of materials to function over their long lifetimes. Consequently, at the initial design stage, it is essential to innovate technology, principles, materials, and use methods to minimize resource consumption. Some corporate actors may remain indifferent to this situation or even exploit it, since selling consumables increases commercial profits. However, corporations with a sustainable development approach will seek win-win solutions for business and the environment, potentially attaining a competitive advantage.

Consumers' habits play an equally important role in the achievement reduction goals. For instance, reducing paper and ink consumption by using digital methods to transmit files, making use of both sides of sheets of paper, and sharing items through exchange service platforms can significantly decrease the use of materials.

1. *Application of innovative technologies and materials:* Applying new technologies and materials can effectively reduce the use of consumables. For example, Dyson vacuum cleaners use cyclone separation technology, eliminating the need for bags and consuming zero materials. This technology not only enhances vacuuming efficiency but also reduces the product's environmental impact, establishes a positive brand image, and is a market success. Additionally, the innovation and application of nanomaterial coatings significantly increased the surface strength and gloss of the vacuum, reducing the frequency of maintenance and use of cleaning agents, thereby extending product lifespan.

2. *Design of small circulation systems:* Designing material flow circulation systems called the cascade method[3] promotes resource recycling. For example, Roca's W + W integrates a washbasin with a toilet, using clean water from the washbasin to flush the toilet, conserving precious water resources. We expect such designs to enjoy increasing popularity.
3. *Optional operating modes or systems:* To meet the diverse needs of users in real-life scenarios, we need to provide functions that use varying amounts of resources to avoid unnecessary waste. One example of this is toilets that have different flush buttons corresponding to the volume of water required for flushing (feces or urine).
4. *Application of intelligent technology:* Use sensors and other technologies to automatically identify and select optimal function modes. This supplementation to intelligent design guidelines allows technology to make choices. Smart technology not only facilitates use, but also effectively saves resources. Examples include dishwashers that detect the number of dishes and how dirty they are to determine washing temperatures and water volume.
5. *Default settings:* Design should compensate for human weakness, as people often do not consciously choose environmentally friendly methods. Therefore, we should configure default settings to the modes that save the most resources. For instance, we should design printers to default to double-sided printing, encouraging users to develop environmentally friendly habits.

7.2 Lowering Energy Consumption

Lowering energy consumption involves decreasing the quantities of energy consumed throughout all stages of a product's lifecycle, with a primary focus on manufacturing, transportation, storage, and use. Decreasing energy consumption is not as intuitive as reducing the consumption of materials. For example, it is difficult to observe the energy-saving effects of LED lights without monitoring one's electricity meter. By contrast, people are quite sensitive to automobile fuel consumption due to the tangible costs of refueling.

Fossil fuels have significantly contributed to the development of modern society, with the result that we rely heavily on non-renewables such as coal, oil, and natural gas. Although the planet's shale oil reserves are vast, we will eventually exhaust them. While nuclear energy technology is developing, it remains controversial. Recently, the rapid development of new clean energy sources has offered hope for alleviating our concerns about diminishing energy resources. Yuval Harari, author of *Sapiens: A Brief History of Humankind*, writes:

---

[3] The cascading approach entails designing a path for material regeneration, recycling materials step by step into a future product or component, meeting increasingly lower demands with each iteration, until the material is no longer useable and transforms into heat energy through incineration, minimizing material waste. See Chap. 4 for details.

> The core of the Industrial Revolution is actually the revolution of energy conversion. We have seen time and time again that the energy we can use is actually infinite. To be more precise, the only limit is our ignorance. Every few decades, we can find new sources of energy, so the total amount of energy that humans can use is actually increasing.[4]

In the future, numerous new forms of energy, such as solar, wind, hydrogen, tidal, and bioenergy, will emerge, providing cleaner and safer options.

Even so, reducing energy consumption remains crucial, since we are still far from efficiently generating inexhaustible quantities of clean energy. Future technology cannot solve today's problems and we must face the current situations of energy shortages and inequality in distribution. Moreover, new energy development is initially costly, preventing widespread benefits. The current scarcity and unfair allocation of energy remain serious global problems.

1. *Product Manufacturing*

We can reduce energy consumption during manufacturing by choosing materials that require little energy to refine and process or by continuously improving energy efficiency through technological innovations. We should opt for efficient, environmentally friendly manufacturing processes. For example, the producing industrial ceramics consumes significant energy, thus requiring new materials and technologies. Using renewable energy sources such as solar, wind, and bioenergy is critical, minimizing fossil energy use. Designers should acquire knowledge about materials, manufacturing processes, and new energy sources, or regularly communicate with engineers to stay informed about technological advances.

2. *Transportation and Storage*

Energy consumption from transportation and storage is cause for increasing concern. Different modes of transportation consume varying amounts of energy. For long-distance transportation, ships consume less energy than trains and cars,[5] while airplanes consume the most energy. Storing physical products incurs energy and space costs, along with management expenses. Storage of non-physical data also comes with huge costs. Data centers are even called "energy-consuming beasts,"[6] a

---

[4]Yuval Noah Harari. *Sapiens: A Brief History of Humankind*[M]. Mcclelland & Stewart, 2014: 343.
"At heart, the industrial revolution has been a revolution in energy conversion. It has demonstrated again and again that there is no limit to the amount of energy at our disposal. Or, more precisely, that the only limit is set by our ignorance. Every few decades we discover a new energy source, so that the sum total of energy at our disposal just keeps growing."

[5]Jia Shunping, Peng Hongqin, Liu Shuang, et al. A review of research on the correlation between transportation and energy consumption[J]. *Transportation System Engineering and Information*, 2009, 9(3): 6–16.

[6]Zhu Yan, Yao Jinnan. How data centers became "energy-consuming giants"[N]. China Energy News, 2019-07-22(1). The article states: "In 2010, the total electricity consumption of data centers in Beijing was 2.66 billion kilowatt-hours. By 2015, this number had increased to 6.7 billion kilowatt-hours, accounting for 7% of the total electricity consumption in Beijing. Last year (2018), it exceeded 10 billion kilowatt-hours."

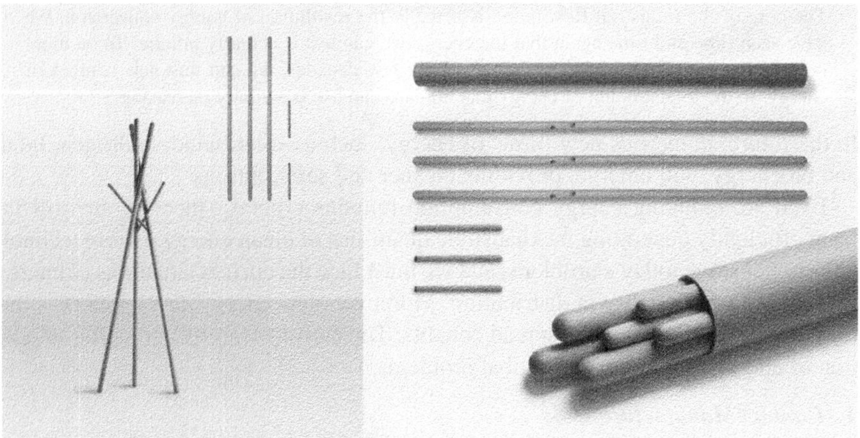

Fig. 7.3 NUDE coat rack made of six sticks, designed by Shen Wenjiao's team (Image source: https://bddwatch.com)

new issue in the information age. The resolution of such problems relies on technological upgrades and improvements in management levels.

However, design measures can reduce energy consumption in storage and transportation. Reducing product and packaging volume and weight, designing compact products with high storage density, and DIY products can improve transportation efficiency and reduce energy consumption. For instance, the NUDE coat rack, assembled using a mortise and tenon structure, consists of only six sticks. Its packaging, a cylindrical paper tube with a 10-centimeter diameter, significantly improves transportation efficiency and saves materials, thus reducing consumption (Fig. 7.3).

Additionally, designing a service system that integrates transportation services into the sales chain, making full use of logistics capacity and avoiding empty runs, can help. For example, some e-commerce companies recycle consumers' old, unused items while making new deliveries. Using a distributed system, such as developing localized production and consumption systems, reduces transportation mileage, thereby lowering energy use and promoting local economic development.

3. *Product Use*

Reducing energy consumption during use requires effort from both users and designers. Users should acquire energy-saving knowledge and develop good habits, such as understanding different lighting technologies' energy-saving qualities to make informed decisions.[7] People's lifestyles and habits significantly impact energy

[7] At present, if LEDs replace traditional incandescent lamps, it will save 90% of electricity consumption. The energy-saving effect of LED bulbs is very clear. They have reduced energy consumption to 1/8 of incandescent lamps and 1/2 of electric energy-saving lamps. Yan Kai, Song Qingjun, Wang Meixia, et al. Analysis of the current status of light efficiency of LED lighting products[J]. *China Lighting Electrical Appliances*, 2016(11): 31–32, 54.

consumption. This includes such things as the frequency with which they run their washing machines, how much laundry they load in each wash, and which washing mode they choose. Additionally, driving habits affect automobile fuel consumption (rapid acceleration and frequent braking increase energy consumption).

Another often overlooked detail is standby power consumption. Household appliances plugged into power sources, even when idle, still consume energy. Although individual appliances' energy consumption is negligible, the cumulative standby power consumption of numerous appliances in multiple households is substantial. In the US, standby power consumption of all appliances is equivalent to the power generation of 18 power plants. The daily power consumption of a microwave oven's clock can exceed the microwave ovens—most household microwaves run fewer than 10 min per day.[8] Chinese households typically have 20 to 30 types of appliances, with routers, refrigerators, air purifiers, and built-in kitchen appliances powered 24 h a day. If every Chinese household were to turn off these appliances when not in use, the saved electricity could supply all the households in three northeastern provinces.[9] Developing the habit of turning off power for idle appliances is a crucial strategy for reducing energy consumption.

Designers should prioritize creating efficient energy-saving products, buildings, and systems that use passive energy, opting for renewable energy sources like solar, wind, and bioenergy. Using human bioenergy, such as electricity generated by treadmills, exercise bicycles, arm-powered watches, and wind-up radios, flashlights, or razors, is also beneficial. We can design smart control systems with motion-sensing road illumination, display-lighting systems, and public escalators to reduce equipment idling and avoid energy waste. We can also design energy-recovery systems, such as those that harness electricity generated by vehicle loads for lighting, as well as user-oriented systems that provide intuitive energy-saving prompts—for example, standby power consumption displays that remind users to switch off devices and reduce energy use. Designing products for collective or shared use, like shared taxis, decreases energy use, reducing overall consumption.

The Google Nest Learning Thermostat is a smart device that automatically learns and adjusts to consumer behavior. It detects when consumers are home or out of the house and acquires information about their preferred temperature settings, adjusting accordingly to meet comfort needs while reducing energy consumption and costs.

[8] Shedroff. N. Design is the problem: The future of design must be sustainable[M]. Liu Xin, Qin Jingyan, trans. Beijing: Tsinghua University Press, 2011: 127.

[9] Source of information: 2017 CCTV public welfare propaganda film "Saving—Appliance Standby" (http://gongyi.cctv.com)

7.3 Harmless Design

Harmless design aims to minimize emissions of toxic waste, garbage, sewage, and harmful gases, reducing environmental impact and harm to humans and animals. Its primary strategy is choosing alternative materials and sources of energy and continuously optimizing clean production processes to reduce toxic and harmful substance use and emissions throughout a product's lifecycle.[10] Additionally, nonmaterial design and localization strategies can reduce transportation costs and carbon emissions.

7.3.1 Toxic and Harmful Substances

Toxic and harmful substances include heavy metals, formaldehyde, chlorinated cleaners and solvents, particulates, hormones, hormone-like solvents, dioxins, and other pollutants. Regulators and companies must strictly control these substances, since consumers cannot identify them. For example, PVC materials may contain toxic plasticizers (such as phthalates[11]). This chemical leaches out during product use and disposal. A carcinogen, it enters the human body through food, air, and skin and disrupts the human endocrine system. Despite this, PVC enjoys immense popularity because it is elastic, transparent, durable, and cheap. It easily transforms into toys, packaging, cables, and other products. With various additives, it turns into a hard plastic, applicable for manufacturing building decorations, floor tiles, pipes, and other materials with a wide range of uses. Although many countries, including those in the European Union, have imposed restrictions on phthalate,[12] there are still products on the market that do not comply with regulations. How to improve the production process of PVC and develop new alternative materials have become

[10] The pollution from electroplating processes is severe, producing large quantities of toxic, harmful wastewater, exhaust gases, and waste residues during pretreatment and production processes. Not only does electroplating pollute the environment, but it also is toxic to organisms, even causing cancer. These toxic substances are difficult to filter out, posing a significant risk to workers' health. Therefore, designers and engineers should avoid choosing such surface treatment processes. Reference: Zou Senlin. Progress in the treatment of electroplating wastewater[J]. Guangdong Chemical Industry, 2010, 37(8): 142–144.

[11] Phthalates can make plastics soft and pliable. Toys containing them may cause abnormal development of reproductive organs and a decrease in sperm count and are considered to have reproductive toxicity and endocrine disrupting properties. Its health risks have raised widespread concern. Currently, countries around the world have begun to take measures to restrict products containing harmful plasticizers such as phthalates. Source: Greenpeace official website: https://www.greenpeace.org.cn

[12] The European Union (EU) implemented restrictive regulations on phthalates in January 2019, specifically requiring that the total concentration of DEHP, DBP, BBP, and DIBP not exceed 0.1%. Since the concentration usually needs to be higher than 10% for plasticizers to soften plastics, this regulation is generally understood as a de facto ban.

pressing issues. The chemical industry has been exploring a series of phthalate-free alternatives. But until new breakthroughs occur, designers are advised to avoid using this material as much as possible.

In addition to the significant toxicity of the chemicals mentioned above, otherwise safe materials that do not degrade for absorption in nature are also toxic. For example, the plastic waste we often mention, including food-grade plastic waste, if dumped in large quantities in nature, causes serious harm to the ecosystem. According UNESCO data, plastic fragments cause the death of over one million seabirds and more than 100,000 marine mammals annually.[13] The excessive use of any material can cause harm to the environment or animals, especially in the context of large-scale industrial production. The ancient Chinese saying, "There is no beauty or ugliness in things, but disaster comes from excess," applies here.

Although people usually feel an affinity to natural materials, some of them are toxic and harmful to humans. These include some wild fungi, materials containing alkaloids, and foods containing aflatoxin, while some natural stones are radioactive. Natural asbestos produces a large amount of lung-disease causing dust during mining and use. Some otherwise harmless natural substances may cause harm due to environmental changes or improper handling. For example, animal (including human) excrement, long an important link in the natural ecological chain as a nutrient source for microorganisms and plants, can become a public health hazard and even cause serious epidemics, if not properly handled in increasingly crowded urban communities.[14]

7.3.2 Alternatives

Alternatives mainly refer to two aspects: materials and energy. First, products are composed of materials, and the choice of these materials and their processing methods determines a significant portion of the product's environmental impact, directly or indirectly affecting our health and ecosystems. Although designers have limited influence on major environmental issues, they do have a certain degree of say in the selection of materials and processing techniques. If designers possess sufficient knowledge of materials and engineering, they may choose materials and

[13] UNESCO. Facts and figures on marine pollution. [eb/ol].[2021-09-16] http://www.unesco.org/new/en/natural-sciences/ioc-oceans/focus-areas/rio-20-ocean/blueprint-for-the-future-we-want/marine-pollution/facts-and-figures-on-marine-pollution/

[14] In the nineteenth century, cholera, a "century disease" that swept across Europe and the rest of the world, broke out four times in the UK on a large scale, causing a large number of deaths and a serious public health crisis. British medical researcher John Snow discovered that the pathogen, Vibrio cholerae, spread through human excrement seepage, drainage pipe leakage, and water source pollution. As a result, beginning in 1859, the UK built a complete urban sewerage system over 6 years, thereby eliminating fecal pollution of water sources and disease transmission. See: Lu Ming, Chen Daijie, Yin Yu. The 1854 London cholera and the father of infectious disease—John Snow[J]. *China Antibiotics Journal*, 2020, 45(4): 347–373.

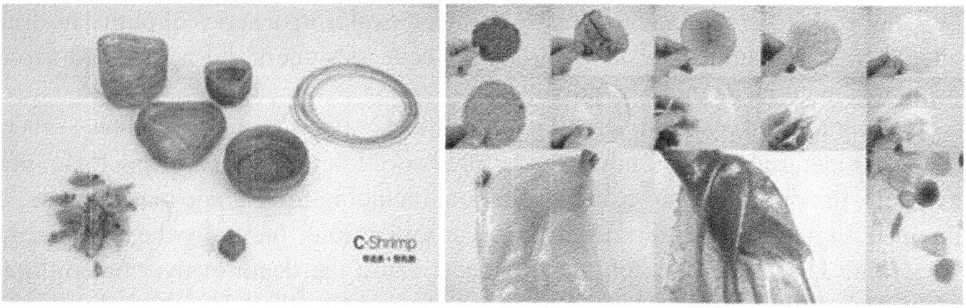

Fig. 7.4 Chitin Material Research (Image source: Huang Sunjie)

manufacturing methods with lower environmental impacts. However, the reality is that most designers are more focused on creating "hot products," emphasizing new selling points and flashy designs, while showing little concern for the source of materials and the fate of products after disposal, as well as a lack of interest in the by-products and pollutants of the manufacturing process. Nonetheless, with the growing attention to environmental issues and companies seeking sustainable development, an increasing number of individuals are beginning to engage in the research and application of new alternative solutions.

The rapid development of biodegradable materials is a typical example. These materials not only conserve valuable non-renewable resources and reduce environmental impact, but they also have the potential to create new trends and styles, presenting new opportunities for businesses that are full of promise. Huang Sunjie, a graduate student at Tsinghua University's Academy of Arts and Design, has attempted to extract chitosan from seafood waste such as crab and shrimp shells, and then produce biodegradable materials. By incorporating various natural additives, these materials can exhibit different characteristics, such as strength, flexibility, and transparency, and can also be made into substrates for 3D printing (see Fig. 7.4).

Creating new degradable materials out of food processing waste kills two birds with one stone. Many people do not know that, from the farm to the pot, 99.8% of coffee beans end up as waste (including fresh pulp and brewed coffee grounds), with people only consuming 0.2%.[15] In designing and developing the OOObject project, Lin Jihua, a designer from Hong Kong, extensively incorporates food waste as a primary material. For example, the Cello bath scrub, made from coffee grounds, walnut shells, lobster shells, pig bristles, and other biodegradable compounds (Fig. 7.5) not only functions well, but can also achieve 100% degradation in soil with the correct temperature and humidity.

The environmental benefits of using biodegradable materials in packaging are evident. Although optimally we would avoid disposable packaging, this is challenging in sectors like the medical and food industries. In these cases, using

[15] Pauli. Blue Economy 2.0: The Latest Report for the Roma Club [M]. Xue Lin, Hu Xilin, Trans. Shanghai: Xuelin Publishing House, 2017: 81.

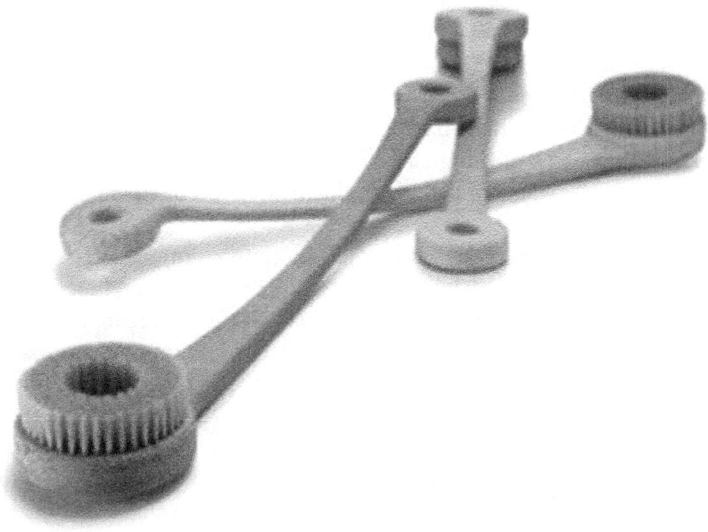

Fig. 7.5 Cello Bath Scrub made from food waste (Image source: Lin Jihua)

biodegradable materials is a prudent choice. Italian designer Emma Sicher integrates fruit and vegetable waste with fermenting bacteria to create disposable packaging. She extracts microbial cellulose from waste, which, after drying, forms a semi-transparent flexible material akin to paper or plastic. Packaging bags made from this material require minimal energy and resources during production and can ultimately become fertilizer, transforming into fertilizer that nourishes the soil after disposal.

There are also familiar items such as the glutinous rice paper in candy packaging, the corn roasting packaging cup outside ice cream, and coffee cups made from natural grains, etc. These packages are both functional and avoid waste generation, while also being part of delicious food. The mycelium material, which will be introduced in Chap. 8 "Design Learning from Nature," is also a potential alternative for future products and packaging materials. Other examples include using glutinous rice paper to package candy, compostable corn plastic ice cream containers, and coffee cups made from natural grains. These packages are functional, minimize waste generation, and can even be consumed as part of the food. Mycelium material, discussed in Chap. 8, "Design Learning from Nature," also holds potential as a future alternative to plastic for products and packaging.

However, we should temper our expectations regarding materials. Even if a material is harmless, producing it can have a negative impact on the environment. For instance, papermaking can cause water pollution, adhesives may contain toxic additives, and bamboo board production consumes energy. Even with strict control over sources, additives, and the production process of ecological materials, improper use and disposal can cause environmental issues. For example, polylactic acid

(PLA), a bio-based material, theoretically decomposes into water, carbon dioxide, and organic matter within 75–80 days under specific conditions of temperature, humidity, and microbial presence, but only degrades naturally with difficulty.[16] Moreover, the use of significant quantities of valuable food crops (such as corn) to make bioplastics is contentious from environmental, social, and economic perspectives.[17] Thus, we must evaluate material substitution solutions with respect to that material's entire lifecycle and include in that evaluation the complete product system design.

On the other hand, there are energy substitution solutions. Renewable energy sources, including solar, wind, hydropower, tidal, hydrogen, and bioenergy, are crucial for reducing pollutants and greenhouse gas emissions. Technological advances and cost reductions will facilitate the widespread adoption of these clean energy sources. For instance, solar and hydrogen-powered vehicles exemplify clean energy applications.

Christoph Behling Design Studio designed the Serpentine, a solar-powered boat operating on the lake in London's Hyde Park since 2006. Powered by 27 diamond-distributed curved solar panels that form the hull's elegant shape, the boat operates quietly without emitting pollutants. Compared to power from traditional fuel, the Serpentine eliminates 4900 pounds (approximately 2223 kilograms) of carbon dioxide each year (Figs. 7.6).

However, alternative energy sources can also have environmental impacts. Although renewable energy is clean during use, producing the equipment used to harness this energy consumes resources and generates emissions. For example, the production of solar panels, including the purification of polycrystalline silicon, can result in toxic emissions and water pollution. Furthermore, producing the auxiliary materials needed for solar panels, such as aluminum alloys, steel, and cables, has an environmental impact.

Electric vehicles, an essential part of national environmental protection strategies, may use energy derived from traditional coal or oil. The energy production process consumes non-renewable resources and produces pollutants. Additionally, the disposal of batteries from electric vehicles poses environmental concerns. Wind

[16] Emadian S M, Onay T T, Demirel B, Biodegradation of Bioplastics in Natural Environments[J]. Waste Management, 2017, 59:526–536.

[17] The production of PLA and other starch-based plastics relies on food crops, including potatoes, corn, sugarcane, and cassava, which may cause deforestation, destruction of natural species habitats, and occupation of agricultural land. Although the plastics industry is researching the use of agricultural waste (corn stalks, sugarcane bagasse, and other lignocellulosic fibers) and waste gas (carbon dioxide, methane) as raw materials, food crops will still serve as the primary raw materials in the short term. According to an estimate by Professor Wen Zongguo of Tsinghua University, the cumulative demand for biodegradable plastics in China will reach 22 million tons over the next 5 years. If we were to produce all PLA using corn starch, the demand for corn for plastic production would reach about 20% of China's total corn output in 2019, occupying 7% of China's grain sowing area. Source: Greenpeace. Decoding "Degradable Plastics": Definition, Production, Application, and Disposal [R/OL]. (2020-12-30) [2021-09-16]. https://www.greenpeace.org.cn/degradable-plastics-report-20201230

Fig. 7.6 Christoph Behling's Serpentine solar-powered boat (Image source: Christoph Behling Design)

power, although less problematic in this regard, still changes surrounding landscapes, creates noise pollution, and can harm migrating birds. The manufacture, transportation, installation, maintenance, and disposal of wind turbines also involve significant costs and material consumption.

Ultimately, it is impossible to eliminate environmental impacts entirely. Our efforts should focus on minimizing unnecessary consumption and waste, while exploring smart system solutions. The goal of finding alternative materials and energy should involve evaluating and selecting options based on their functional units throughout their lifecycles. Additionally, we must beware of being misled by concepts labeled as natural, green, or ecological. In some cases, plastics like PET or PE-HD may be more sustainable if they significantly extend product life, provide services to more people, and are easy to recycle.

7.3.3 Dematerialization Design

The ultimate way to reduce the consumption of resources is non-material design, leveraging information technology (digitization) or service methods to replace traditional material products. Since the 1990s, with the advent of digital technology, dematerialization has transitioned from a concept to a practical approach. Economic development has shifted from relying on the production and sales of material products to offering valuable information and services to consumers.

Digital technology has significantly transformed both the world and how people live. Examples include online music, e-books, and magazines, emails, and social media. These innovations have made traditional forms of communication nearly obsolete. The new generation of digital natives is used to these technologies, like fish in water.

Effective use of intelligent digital technology can help us achieve our environmental sustainability goals. For instance, artificial intelligence (AI) can assist in design, meeting user needs, reducing the waste trial and error generate, and improving utilization of resources. 3D printing technology supports personalized customization and reduces waste. It enables local production and distribution through digital file transmission, obviating the carbon emissions logistics transportation creates. However, digitization does not inherently reduce environmental impact. While digital offices increase the efficiency of information sharing and reduce paper and energy waste, they can also lead to increased paper consumption if the people working in them find printing digital documents convenient. Therefore, environmental awareness among office workers is crucial.

Service design is another vital approach to meeting user needs non-materially. Consumers fundamentally seek functions rather than tangible, resource-consuming products. For instance, people need music and entertainment, not necessarily records or tapes; they need reading material and intellectual nourishment, not necessarily paper books and magazines made from paper; they need transportation, but not necessarily private cars. Service design, combined with digital technology, can provide these functions, meeting user needs through rental services and shared transportation, for example, without selling physical goods. For more on product service system design, see Chap. 9.

7.3.4 Localization Strategy

Despite the effectiveness of digitization and services when it comes to reducing resource consumption, often we cannot avoid the need for material products. That is why localization strategies have value. Localization is the opposite of globalization, which surged in the latter half of the twentieth century due to free trade and information technology. In 2005, American political commentator Thomas L. Friedman's *The World Is Flat: A Brief History of the Twenty-first Century* came out. Although Friedman notes challenges, such as infectious disease and terrorism, he overall depicts the future of globalism optimistically, characterizing it as having complementary advantages and promoting global unity.

However, due to scrutiny concerning its effects on the environment and consequently people, globalization has slowed and localization has gained acceptance. *There is a popular view that the closer the origin of the food you buy is to where you live, the more environmentally friendly it is. In this scenario, growing your own food is the best option.* The CSA[18] farm model, which became popular in China at the

[18]CSA: community-supported agriculture. *Community-supported agriculture* (*CSA model*) or *cropsharing* is a system that connects producers and consumers within the food system closer by allowing the consumer to subscribe to the harvest of a certain farm or group of farms. It is an alternative socioeconomic model of agriculture and food distribution that allows the producer and

beginning of the twenty-first century, is a counter-response to large-scale industrialized agriculture, with many city dwellers joining CSAs.

Localization is appealing for two reasons: proximity to a product's source reduces the need for long-distance transportation, saving energy and reducing emissions, and supporting local businesses keeps profits in the community. This applies not only to food production but also to the processing and manufacturing of other products and energy supplies. Local distributed energy systems, such as solar, wind, and hydroelectric power, can reduce the environmental impact of centralized power supply during transmission (see Chap. 11 for related discussions).

However, localization is not without problems. It still requires comprehensive consideration of the environment. Distance affects energy consumption and carbon emissions during distribution, service, and recycling, but it is not the sole factor we need to consider. Not all products are suitable to localization strategies. Global industrial chains are complex and different regions have varying suitability for producing certain goods due to environmental factors. For example, water-scarce African regions are inappropriate for rice cultivation, Nordic regions for olive cultivation, and regions like Inner Mongolia and Xinjiang for grazing when compared to the Jiangnan Region. Localization can result in higher costs and carbon footprints for some products. Large-scale production near raw material sources with abundant labor often results in increased resource efficiency. Thus, localization should be seen as a supplementary approach to globalization, not a wholesale replacement.

7.4 Reflection on Design for Reduction

The Design for Reduction strategies discussed above focus on minimizing material and energy consumption, reducing toxic and harmful emissions, choosing alternative solutions, and employing digital, service, and localization methods. These technical measures and suggestions are meaningful but not sufficient for sustainable design. Reduction is foundational to many design strategies and linked to social and economic efforts to create a responsible, sustainable society. Achieving this requires collaboration among governments, enterprises, and consumers. Corporate development models must transform alongside policy norms and technical advances. We often criticize retail therapy[19] and the massive waste caused by consumers' prefer-

consumer to share the risks of farming. (https://en.wikipedia.org/wiki/Community-supported_agriculture)

[19] Retail therapy refers to shopping with the intention to improve one's mood or mental state. Shopping can fill certain emotional voids in a person's life. The impulse to shop is to some extent a survival instinct. It dates back to when humans lived by hunting and gathering: people would grab what they wanted when they saw it, even if they did not need it because it was likely they would not come across it again. The items purchased during retail therapy are called "comfort buys." Commercial design can effectively stimulate and encourage this unconscious impulsive consumption behavior.

ence for novelty over old things, the result of deliberate corporate marketing techniques. *Companies use creatively ingenious and highly engaging product advertisements to subtly implant commercial goals into consumers' minds, transforming them into their inner desires. While consumers express their "self" through shopping, they are, in fact, fulfilling the logic of capital and the aspirations of businesses. In this cycle of continuous buying and disposing, a significant amount of valuable resources is consumed, leading to environmental pollution.* As Professor Liu Guanzhong of the Academy of Arts & Design at Tsinghua University once said, "fashion is short-lived." Think of all the discarded things you have in your home: clothes, fitness equipment, toys, appliances, and the unused kitchen utensils, the juicers, yogurt makers, pasta makers, bread makers, rice cookers, and so on. Momentarily popular, these items quickly gather dust and are forgotten. Traditional business models rely on continuous consumption of material products, which is unsustainable due to outsized resource consumption and pollution. How to transform this model, reduce dependence on physical products, and still ensure that companies have room for sustainable development and product service system design is an important direction for transformation (see Chap. 9).

Consumers must mature and awaken to this reality. We are rational animals capable of long-term planning and reflection. Our preference for high-quality, simple lifestyles, presents both new challenges and opportunities for designers. Beyond fast-moving consumer goods, designers can provide durable products, services, experiences, emotions, stories, and dreams.

"Reduction" does not mean scarcity or superficiality; rather, it should represent a higher quality and a more elevated approach to development and lifestyle. Fundamentally, this signifies a thorough reflection and transformation of the traditional industrial development model characterized by "mass production, mass consumption, and mass disposal." Relying solely on stimulating the consumption of material products and pursuing extensive development is unwise, as resource consumption and pollutant emissions will inevitably occur simultaneously. Only by reducing at the source and transforming development models can we hope to approach the ideal of a sustainable society.

Chapter 8
Design Learning from Nature

Abstract Nature has shaped the Earth into its present form through 3.8 billion years of evolution. This process has involved countless instances of "design," correction, selection, and elimination. Although humanity has yet to grasp the essence of nature, we can still gain insights into the fundamental laws and operational logic of natural creation, such as "adaptation," "efficiency," and "interdependence." Undoubtedly, if we are to learn about sustainable design, nature is our best teacher. The imitation and study of nature in terms of materials, forms, structures, principles, and systems are the explorations and practices that designers undertake under this philosophy.

Keywords Bionics · Biomimetics · Biomimicry · Nature-Inspired Design

As Leonardo da Vinci observed in the Codex Atlanticus, "Everything comes from everything; everything is made by everything; everything transforms into everything because everything that exists in the elements is also composed of these elements."[1]

After 3.8 billion years of evolution, nature has shaped the Earth into what it is today. This has entailed countless designs, revisions, selections, and deletions. Faced with various extreme challenges, such as variations in environments differences and temporal and spatial change, nature has provided exquisite solutions. Its grand evolutionary project continues unabated. Although humans cannot yet fully comprehend the depth and breadth of nature, we can still observe the fundamental laws and the operational logic of natural creation, such as adaptation, efficiency, and interdependence.[2] Nature knows better than any of us what design is the most

[1]Excerpt from the preface to Gunter Pauli's book, *Blue Economy 2.0* (2017), written by Ashok Khosla.

[2]Myers W. *Bio Design-Nature·Science·Creativity* [M]. London: Thames & Hudson Ltd., 2018: 10–125.

X. Liu et al., *Design for Sustainability*, Sustainability and Interdisciplinarity,
https://doi.org/10.1007/978-981-96-8657-5_8

reliable, most appropriate, and most lasting.[3] *When seeking to create sustainable design, nature is our best teacher.* Imitating and learning from nature to choose materials, forms, structures, principles, and systems is what sustainable designers explore and practice.

The Earth is a closed system, which is why scientists and environmentalists emphasize the preciousness of limited resources, highlighting their importance to human survival and development.[4] Nature wastes nothing, in stark contrast to our man-made world, which often violates natural laws, causing ecological disasters and environmental crises. In the twenty-first century, advances in science and technology have allowed humans to further explore the mysteries of nature and learn from its wisdom. We have observed how nature makes use of minimal resources through simple and ingenious design or combinations to achieve advanced, complex systems and structures, forming interdependent relationships and closed-loop network systems with feedback mechanisms. This understanding has inspired a vast number of innovative design practices, attracting researchers, designers, engineers, and entrepreneurs hoping to find answers to severe problems and pathways to a sustainable future.

This chapter outlines the processes whereby we learn from nature in order to inform efforts at creating sustainable design. It introduces various concepts, methods, and representative design cases related to the concept known as biomimicry design.

8.1 Natural History in Design

Human inquiry into nature's form, harmony, and beauty date back to Ancient Rome, as evidenced by Marcus Vitruvius Pollio's *Ten Books on Architecture* (DE ARCHITECTURA).[5] Pollio's work not only summarizes the proportions and natural laws of the human body, but also applies these to architecture for the very first time. Centuries later, in *The Metamorphosis of Plants*, the poet Johann Wolfgang von Goethe echoes Pollio's curiosity, describing the various forms of plants over

---

[3] Benyus, J. M., *Biomimicry: Innovation Inspired By Nature* [M]. New York: William Morrow & Co, 1997.

[4] Kapsali V. *Biomimetics For Designers: Applying Nature's Processes And Materials In The Real World*[M]. London: Thames & Hudson Ltd., 2016: 8–217.

[5] Marcus Vitruvius Pollio wrote *Ten Books on Architecture* (De Architectura) at the end of the first century BC. Pollio lays out three main standards for architectural design: durability, utility, and beauty (*firmitas, utilitas, venustas*). He proposes that architecture imitates nature: just as birds and bees build nests, humans use natural materials to construct buildings to protect themselves. He also states that the Doric, Ionic, and Corinthian orders were invented to make architecture beautiful and must be designed and built according to the most beautiful human proportions.

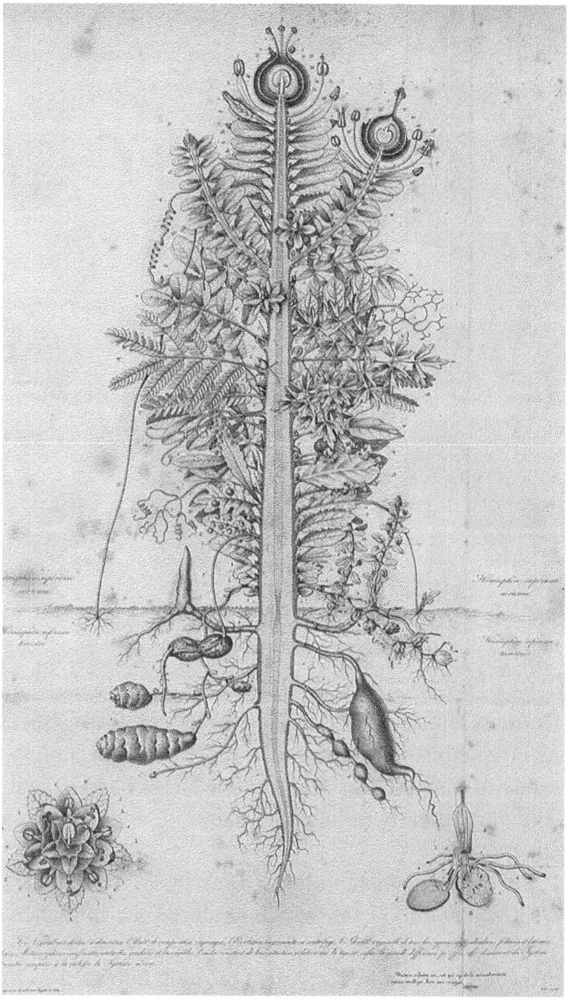

Fig. 8.1 Plant prototype: Goethe asked botanist Pierre Jean Turpin to draw an imagined composite plant (Image source: https://de.wikipedia.org/wiki/Urpflanze)

time, attempting to summarize the essential laws of plant growth and change, proposing that "All shapes are similar and none is like the other."[6] (Fig. 8.1).

Humans' observation and imitation of nature peaked in the late nineteenth century with the Art Nouveau movement. Designers immersed themselves in the smooth lines and patterns of natural flowers and plants and applied them to architecture, furniture, jewelry, clothing, book-binding, and designing other material objects. Many artists of the Art Nouveau movement were inspired by the work of

[6]Goethe, J. W., Gedichte. Die metamorphose Der Pflanzen. Friedrich Schiller: Musen-Almanach Für Das Jahr 1799, Tübingen. S. 17–23. Original German text: All shapes are similar and none is like the other.

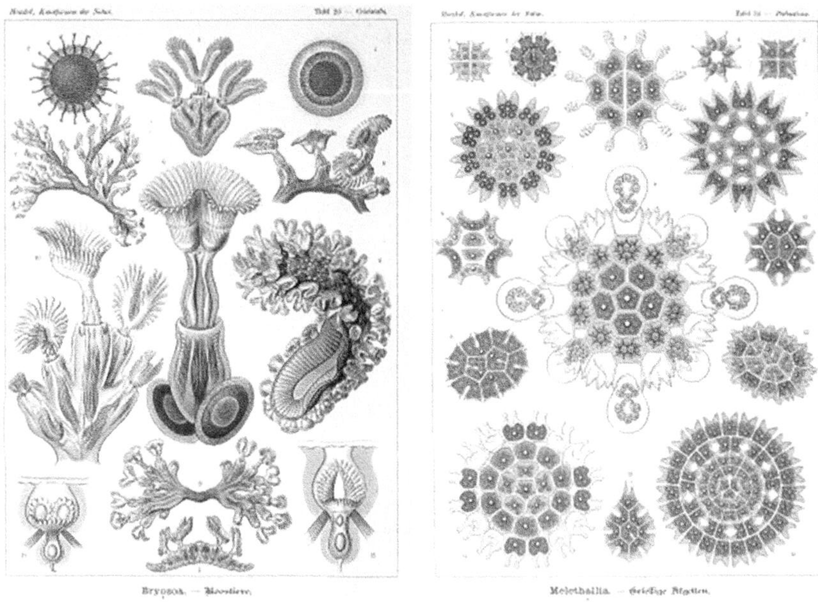

Fig. 8.2 Illustration drawn by Ernst Haeckel

German naturalist and biology pioneer Ernst Haeckel.[7] His book *Art Forms in Nature*, published at the end of the nineteenth century, meticulously described thousands of new species and included 100 hand-drawn images of various organisms (Fig. 8.2), greatly influencing painting, architecture, and design in the early twentieth century.[8]

In 1917, Scottish biologist D'Arcy W. Thompson published the groundbreaking *On Growth and Form*, which explored the influence of the laws of physics on the shape of organisms and developed a new method for studying their evolution and growth. He examined why fish develop into various forms according to their growth environments, which included coping with water pressure, escaping predators, and saving energy.[9] The forms species take result from natural selection, perfectly

[7] Ernst Heinrich Philipp August Haeckel, a German biologist, philosopher, and artist, published the illustrated *Art Forms in Nature* (*Kunstformen der Natur*) at the end of the nineteenth century, which influenced many artists of the Art Nouveau movement.

[8] Myers, W. *Bio Design: Nature, Science, Creativity* [M]. London: Thames & Hudson, 2018.

[9] *Growth and Form* uses mathematical language and physics to explicate the reasons biological forms assume particular shapes. The author believes that gravity plays a decisive role in the shape of organisms. However, if the effects of physical force on biological forms is not a factor, organisms choose "ideal geometry" to solve their morphological problems. In these cases, biological form becomes a mathematical problem, choosing the most effective shape possible while maintaining its original function. Reference: Fournier. *When nature inspires technology* [M]. Pan Wenzhu, trans. Nanchang: Jiangxi People's Publishing House, 2017: 7–16.

Fig. 8.3 The Nakagin Capsule, the most famous work of Japanese metabolism architecture, *Photo by Tomio Ohashi* (Image source: https://sabukaru.online)

combining form and function and adhering to the rules and laws of geometry,[10] such as the logarithmic spiral of mollusks, the Fibonacci (golden ratio) sequence, and the Tyson polygon.

In the early twentieth century, architectural masters like Frank Lloyd Wright, Alvar Aalto, and Ludwig Mies van der Rohe consistently drew inspiration from nature. They focused on integrating the design of indoor and outdoor spaces, using natural materials, and the expression of structure, viewing architecture as part of a larger whole that serves to create the environment. After World War II, Japan also adopted nature-inspired design on a large scale. Designers like Kenzo Tange believed cities and buildings should undergo dynamic growth processes like biological organisms.[11] The Nakagin Capsule Tower, which Kisho Kurokawa designed, is a famous example of metabolism architecture, with 140 precast architectural modules, each independent and replaceable (Fig. 8.3).[12]

---

[10] The mathematical basis for Fractal Theory is fractal geometry, which derives fractal information, fractal design, and fractal art.

[11] Metabolism is a school of thought in the architectural world that emerged in Japan in the 1960s. Metabolists believe that everything in the universe undergoes a metabolic process and our task is to promote the realization of this metabolism. They issued the Metabolism 1960 Declaration, which emphasizes the growth, change, and decay of things, strongly advocates for the use of new technologies to solve problems, and oppose the former view that cities and buildings evolve fixedly and naturally. Metabolists believe that time factors should be introduced into cities and buildings, and the cycles of various elements should be clarified.

[12] Kisho Kurokawa was a student of Kenzo Tange and thus deeply influenced by his design ideas. The Nakagin Capsule Tower he designed is the most famous representative work of Metabolism architecture. The building is made up of a total of 140 stacked capsules that rotate around the cen-

Following the environmental movement of the 1960s and the energy crisis of the 1970s, people's understanding of the environmental impact caused by artificial buildings and industrial products deepened. This gave birth to ideas that embraced adapting to nature and protecting the environment. In the 1960s, bionics emerged as a discipline. As the result of technological development and expanding application fields, different schools of thought formed, gradually creating a series of design concepts that mimic nature. In 1974, biologist Barry Commoner linked nature, humans, and technology in his book, *The Closing Circle*. Commoner analyzed the root of the environmental crisis from an ecological perspective, revealing the brutal effects of technological developments on the natural ecosphere, and proposed the four laws of ecology.[13] Commoner is considered the father of the modern environmental movement.[14]

In the 1970s, Australian ecologists Bill Mollison and David Holgren proposed the concept of permaculture, a complete life design system that supports sustainable development of the earth's ecology.[15] The term permaculture[16] combines the words permanent, agriculture, and culture. Unlike traditional agriculture, permaculture employs systems theory. Using agricultural technology, it focuses on the characteristics and potential connections of the various elements in a system, combining technology, architecture, fauna, flora, and other factors into a self-evolving energy cycle system that provides a sustainable environment for all life to grow.[17] First applied to agriculture, it has expanded into design and culture. Permaculture design focuses on creating a sustainable environment for all life growth through cooperation with nature rather than confrontation.

In a 1989 *Scientific American* article, Robert Frosch and Nicholas Gallopoulos of General Motors proposed incorporating the term industrial ecology into industrial design. Their core view is that industrial processes can mimic ecosystems, in which each type of waste becomes the raw material of another process. Lifecycle analysis is also an important method, with the circular system view of industrial ecology profoundly influencing future generations.

---

tral core at different angles and is 14 stories high. Each unit is connected to the concrete central core structure with only four high-tension screws, allowing units to be replaced. Each capsule measures 4 meters by 2.5 meters, providing enough living space for one person to live comfortably.

[13] The four laws of ecology are: everything is connected to everything else; everything must go somewhere; nature knows best; there is no such thing as a free lunch. Commoner, B. *the closing circle: nature, man, and technology* [M]. Random House Inc. (T), 1971.

[14] Krier. J E. The Political Economy of Barry Commoner[J]. *Environmental Law*, 1990, 20: 11–33.

[15] See the Permaculture network course: https://www.discoverpermaculture.com/permaculture-masterclass-video

[16] Permaculture is a compound word, translated into Mandarin as "sustainable farming" or "sustainable culture." https://www.permaculture.co.uk/what-is-permaculture/

[17] Hu Wenxue, Zhu Runyun, Zhou Jianwen, et al. "Near" Nature and "Respect" Nature—The Practice and Discussion of the Sustainable Development Concept of Pu Men in the Rural Construction of Da Mo Yu Village in Yunnan [J]. *Journal of Yunnan Agricultural University* (Social Science Edition), 2019, 13(3): 88–93.

In 1997, Janine Benyus published *Biomimicry: Innovation Inspired by Nature*, in which she emphasizes that humans should learn from nature instead of extract from it. The biomimicry theory and methods Benyus propose significantly impacted sustainable design innovations. Architect William McDonough and chemist Michael Braungart furthered these ideas in *Cradle to Cradle: Remaking the Way We Make Things* (2002). Using the growth patterns of cherry trees as an example, McDonough and Braungart explain that everything in the world is a symbiotic state of mutual nourishment and dependence, providing a path for the development of a circular economy in architecture and industry.[18]

Nature does not exist merely to provide resources for humans, but should become a model from which we learn. Discovering and revealing the laws and rules of natural ecosystems may inspire us to move toward a sustainable new world. The evolution of human civilization has always entailed learning from nature. With the development of digital and intelligent technologies, the depth and breadth of learning from nature will greatly expand.

8.2 Design Strategies Learned from Nature

Throughout history, there are countless examples of humans drawing design inspiration from nature. For instance, sawtooth grass inspired Lu Ban to invent the saw; the streamlined shapes of birds and their wings inspired airplane designs; the hook structure of burdock seeds inspired Velcro; lotus leaf structure inspired hydrophobic materials; gecko feet structure inspired Professor Andre Geim's team to invent adhesive tape. Then, there are examples like the Living Root Bridge in Meghalaya, India, which grows stronger over time and merges with nature (Fig. 8.4). Bird forms inspired Leonardo da Vinci's flying machine (Fig. 8.5).[19] In fact, imitating nature or drawing inspiration from natural analogies is not a new method.

Learning from nature is interdisciplinary, connecting science, technology, and design and will help solve the problems we face in the twenty-first century.[20] Over the past few decades, advances in science and technology have facilitated

[18]Also known as the living bridge, it has enough strength and durability to minimally disturbing nature. The tree bridge not only grows over time, but also becomes stronger. Even architects marvel at it as an engineering wonder.

[19]The *Codex Atlanticus* (*Codice Atlantico*) is the largest of Leonardo da Vinci's many manuscript collections, comprising 12 volumes and 1119 pages, written from 1478 to 1519. It covers a wide range of topics, including flight, weapons, and botany. Da Vinci believed that birds are machines that work on mathematical principles, and humans have the ability to replicate all the movements of these machines. Da Vinci conducted detailed research on bird flight and designed a flying machine powered by the pilot called the flapping wing aircraft (see Figure 8.5).

[20]Conrad F, Lofthouse V, Escobar-Tello C. Design based on nature—a literature investigation[c]// proceedings of the 18th international conference on engineering and product design education. Aalborg, Denmark: institution of engineering designers, 2016: 8–13.

Fig. 8.4 The Living Root Bridge of Meghalaya, India (Image source: https://en.wikipedia.org/wiki/Living_root_bridge)

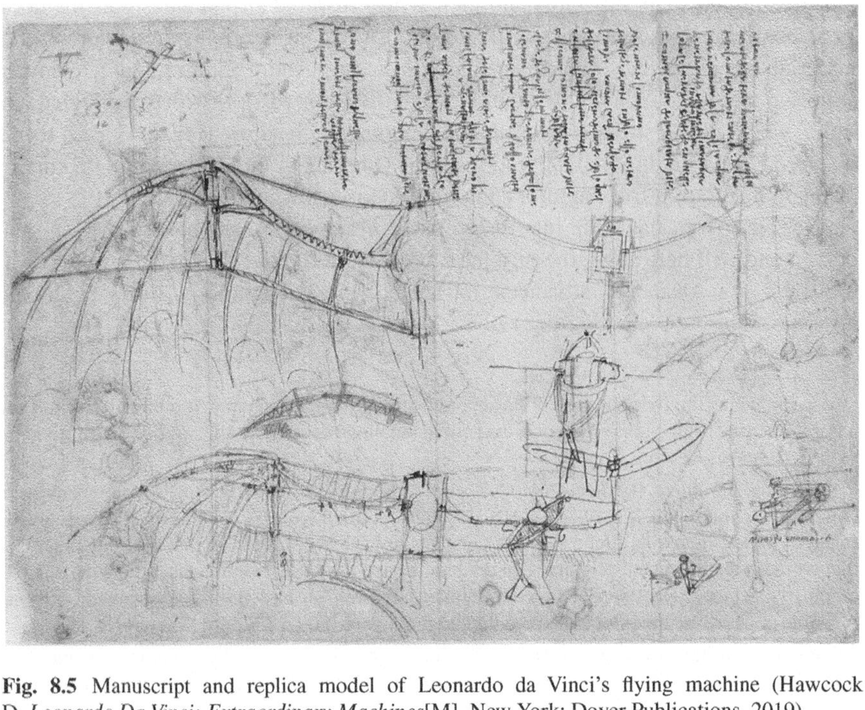

Fig. 8.5 Manuscript and replica model of Leonardo da Vinci's flying machine (Hawcock D. *Leonardo Da Vinci: Extraordinary Machines*[M]. New York: Dover Publications, 2019)

collaboration between life sciences and design, giving rise to biomimetic design concepts with profound impacts on design thinking and methods.

8.2.1 Bionics and Bionics Design

Bionics, simply understood as "copying nature or getting good ideas from nature,"[21] emerged as a discipline[22] in the 1960s when Jack Ellwood Steel formally proposed it at the first Bionics Symposium.[23] A technical science that studies the structure, characteristics, principles, behaviors, and interactions of biological systems, it provides new design ideas, principles, and system compositions for engineering technology. It is also an interdisciplinary subject of life sciences, material sciences, mathematics and mechanics, information sciences, engineering technology, and system sciences.[24] Bionics research includes mechanical, molecular, information and control, and energy bionics. Broadly speaking, we can term all behaviors that imitate animals, plants, minerals, and natural phenomena bionics design. Bionics design applies bionics methods and techniques, yielding with systematic research ideas and solid discipline support.

8.2.2 Biomimetics

In 1958, Otto Herbert Schmitt proposed the concept of *biomimetics*,[25] defined as the study of the formation, structure, or function of biologically produced substances and materials, as well as that of biological mechanisms and processes, especially for the purpose of synthesizing similar products using artificial mechanisms that mimic natural ones.[26] Research in biomimetics is expanding rapidly in three areas. The first is functional morphology, primarily the relationship between biological morphology, structure, and functions, using artificial materials to replace biological materials while maintaining the essence of their functions. Successful examples in this area are mostly related to fluid mechanics, such as designing aircraft wings by imitating the cross-sectional shape of bird wings. The second is imitating the signal and information-processing mechanisms of organisms, which involves bio-computing,

[21] Same as 21.

[22] Bionics is a combination of the words Biology and Technics.

[23] The first Bionics Symposium was held at the Wright-Patterson Air Force Base in Ohio in 1960.

[24] Lu Yongxiang. The Meaning and Development of Bionics [J]. *Chinese Scientist*, 2004(4): 22–24.

[25] The terms Biomimetics and Biomimicry come from the ancient Greek BΊΟΣ Along with MΊΜΗΣΙΣ, BΊΟΣ (BIOS) means life, while MΊΜΗΣΙΣ (IMITATION) means imitation or copying. Due to the same etymology of the two words, people sometimes use them interchangeably. However, the purpose of biomimicry is to imitate nature is to seek more sustainable solutions.

[26] Helfman C Y, Reich Y. *Biomimetic Design Method For Innovation And Sustainability* [M]. Switzerland: Springer International Publishing, 2016: 6–17.

bio-inspired robots, and biomimetic sensor design. The third and newest research area is nano-biomimetics, molecular self-assembly and nanotechnology, including studying the microscopic scale-like structure of shark skin, discovering its antibacterial surface mechanism, and then designing applications.[27]

8.2.3 Biomimicry

Janine Benyus popularized *biomimicry* in her renowned book in the late 1990s. This involves seeking sustainable design solutions by learning from and mimicking the forms, processes, and ecosystems of nature. In it, Benyus proposed that nature is a vast design laboratory that has achieved a perfect ecosystem. Humans should regard nature as a model, mentor, and judge.[28] The more our human world functions like the natural world, the more likely we are to endure.[29] Biomimicry involves understanding natural solutions to contemporary challenges, judging human innovation through ecological standards, and emphasizing the need to learn from nature rather than merely extracting from it.[30] The mission of biomimicry is to imitate the solutions of nature to solve contemporary challenges.[31]

AskNature, an open-source database the Biomimicry Institute launched in 2008,[32] categorizes biological principles by function, providing valuable research resources for scientists, engineers, and designers. Biomimicry design can start from biological research (solution-driven design) or design problems and challenges (problem-driven design).[33]

The first approach starts with biological research, meaning that one first identifies a biological solution and then applies a similar method to address real-world problems. As an example of solution-driven design, researchers found that lotus leaves have a self-cleaning ability that allows them to remain pristine even when covered in mud. The lotus effect[34] occurs is because the surface of lotus leaves have a layer of fuzz and tiny waxy particles that prevent water droplets from sticking to the leaf, which carry away any dust on the leaf surface as they roll off. Using this

---

27 Same as 26

28 Benyus. *Biomimicry: innovation inspired by nature* [M]. Translated by Zhang Zhaofei. Taipei: Walnut Wood Publishing, 2003: 26.

29 De Pauw I, Kandachar P, Karana E, et al. Nature Inspired Design: Strategies Towards Sustainability[J]. *International Journal of Sustainable Engineering*, 2015, 8(1): 5–13.

30 Cen Haitang, Chen Wuyi. The Concept and Evolution of Bionics [J]. *Mechanical Design*, 2007(7): 1–2,66.

31 Helfman C Y, Reich Y. *Biomimetic Design Method For Innovation And Sustainability* [M]. Switzerland: Springer International Publishing, 2016: 6–17.

32 The website for "Ask Nature" is: https://asknature.org

33 Same as 31.

34 The lotus effect refers to the superhydrophobic and self-cleaning characteristics of the surface of lotus leaves.

biological principle, humans invented hydrophobic fabrics with excellent anti-fouling performance, used in various clothing and fabric designs.

The second approach begins with design problems and challenges, followed by searching for potential biological solutions. Japan's Shinkansen train provides an example of problem-driven design. Because the train travels at immense speeds, whenever it exits a tunnel, compressed sound waves release in a sonic boom, disturbing the surrounding residents. When the designers redesigned the train head, they decided to look to nature for answers. They found that when a kingfisher dives from the air into water, there is hardly any resulting splash because its beak and body shape adhere to superior fluid dynamics. The redesigned train is quieter, faster, and more energy efficient than before.

If we invest enough time energy, we find that nature has provided solutions to almost all our design problems. The AskNature database contains large quantities of practical biological knowledge, effectively helping designers uncover the mysteries of and wisdom contained in nature and transform them into real-world solutions.

From bionics to biomimetics and biomimicry, humanity continually deepens its imitation of and learning from nature—evolving from the replication of simple forms to emphasizing interdisciplinary, integrated research, ultimately focusing on innovation and sustainability. The development direction of this field manifests in leaps forward in people's understanding. Relatively speaking, bionics and biomimetics focus on technical solutions applied to industries with high technical complexity, while biomimicry involves learning from nature to seek innovative sustainable solutions, largely in areas with low technical complexity.[35]

8.2.4 *Nature-Inspired Design*

The Nature-Inspired Design (NID) strategy a team at the Delft University of Technology in the Netherlands proposed builds on biomimicry and cradle-to-cradle approaches. NID involves solving human issues by studying the forms, structures, systems, and strategies of organisms, differentiating from purely biological replication. In 2010, the team at Delft presented a paper entitled, "Nature-Inspired Design: Strategies Towards Sustainability" at the Knowledge Cooperation and Learning for Sustainable Innovation ERSCP-EMSU[36] Conference, pointing out that nature-inspired design strategies stem from "learning from nature" and regard nature as a design strategy for sustainable development paradigms.

In 2015, the team published the *Nature-Inspired Design Handbook*, providing a set of principles and tools for cross-disciplinary collaborative design sourced from nature. The handbook notes that design solutions should have a positive impact, rather than just passively reducing ecological impacts and economic costs. Simply

---

[35] A sonic boom is a physical phenomenon that occurs when an object moving in the air breaks the sound barrier, producing a shock wave accompanied by a loud noise that sounds like an explosion. The energy a sonic boom creates is enormous.

[36] ERSCP:the European Roundtable of Sustainable Consumption and Production; EMSU: the Environmental Management for Sustainable Universities.

put, the goal of design should be added value.[37] The NID handbook lays out six important design principles[38]:

1. Waste equals food;
2. Use renewable energy;
3. Be locally attuned and responsive;
4. Adapt and evolve to changing conditions;
5. Integrate development with growth;
6. Be resource efficient.

In what follows, we use the life system of an oak tree to interpret the NID handbook principles (see Fig. 8.6). The lifespan of an oak tree is long, up to 300 years,

Fig. 8.6 NID principles in oak trees (source: *Nature-Inspired Design Handbook*)

[37] Iouguina A, Dawson J W, Hallgrimsson B, et al. Biologically informed disciplines: A comparative analysis of bionics, biomimetics, biomimicry, and bio-inspiration among others[j]. *International Journal of Design & Nature and Ecodynamics*, 2014, 9(3): 197–205.

[38] Tempelman E, Van Der Grinten B, Mul E-J, et al. *Nature Inspired Design: A Practical Guide Towards Positive Impact Products*[R/OL]. The Netherlands: Tu Delft, Boekengilde, Enschede, 2015. [2020-03-15]. http://www.natureinspireddesign.nl/handbook.html

with some living up to 2000 years, although that is extremely rare. As part of a forest, when an oak tree sheds its leaves, dies, or decays, it provides nutrients for surrounding organisms and fungi, so its waste becomes the nutrients and food for other organisms *(Principle 1)*. While alive, an oak tree converts sunlight into sugar and carbon dioxide into oxygen, and grows into sturdy wood, using renewable energy *(Principle 2)*. Due to the various scales of and conditions in forests, species must find different ways to depend on each other and prosper together in their environments. The oak tree passes its energy (especially sugar) on to fungi, insects, birds, and other animals, and benefits from the nutrients its biological system feeds it *(Principle 3)*. To survive, an oak tree must adapt to local conditions. Oak trees have a variety of forms and are distributed in different areas of a forest. Whether the onslaught is from storms, severe cold, or disease, some oaks survive, thereby preserving the resilience of the entire forest to external changes *(Principle 4)*. In an oak tree, there is a typical structure: a bottom-up honeycomb unit,[39] modularization, and orderly hierarchy in response to its environment composed of micro units, which self-organize into larger structures according to growth rate and scale. Therefore, oak trees integrate development and growth *(Principle 5)*. Finally, when an oak fully uses these principles, it outgrows other trees by its improving resource utilization *(Principle 6)*.

8.2.5 *Bio-Design*

Bio-design, or bio-inspired design, combines biology and design by integrating biological and biotechnological principles into design practice and incorporating living organisms into objects and structures. This forward-looking, exploratory concept represents a radical approach to design.[40] In his 2012 book, *Bio-Design: Nature, Science, Creativity*, William Myers writes, "One of my goals is to completely and thoroughly design a new life form." Bio-design explores new possibilities and potential solutions, often through experimental or reflective projects that are evocative of science fiction. Some projects resemble art experiments and philosophical inquiries, offering profound inspiration and inner reflection.

In 2019, London's Central Saint Martin's College of Art and Design established a master's program in bio-design[41] that emphasizes interdisciplinary research. The program also seeks to create new design themes that include redefining the use of

[39] The Nature Inspired Design team at Delft University of Technology proposed the NID principles in 2015.

[40] The interior of an oak tree is filled with honeycomb-like structures that are full of air, giving oak good elasticity.

[41] Conrad F, Lofthouse V, Escobar-Tello C. Design based on nature—a literature investigation[c]. Denmark: international conference on engineering and product design education, September 2016: 8–13.

energy, water, air, waste, and materials to promote the circular economy and sustainable development.

Designer Amy Congdon's Biological Atelier project explores the revolutionary changes biotechnology could bring to textile design. Congdon used digital embroidery and 3D printing to print cells onto textiles, creating living material forms. She speculates that in the future we will cultivate clothing instead of manufacturing, with cells shaping luxury fabrics. Her project not only generates innovative designs, but also raises questions for the fashion industry: What would the world look like if we could cultivate ethical, victimless ivory and cross-species fur? What kinds of hybrid textile raw materials could we practically apply? How can an industry driven by novelty, pleasure, and change adapt to a future of increasingly depleted resources?

Although these concepts generate various backgrounds, disciplinary categories, and research, they share three commonalities: First, they all draw inspiration from nature to address social issues or to pose questions. Second, their research is cross-disciplinary, demonstrating the close connection between science, technology, and design. Third, the degree of intersection and commonality in their proposed design principles is significant, with the four laws of ecology influencing the core content of their design methods. Most strategies advocate for renewable energy use, transforming waste into raw materials, interdependence between individuals and systems, and closed-loop relationships. Learning from nature's various design concepts is as idealistic as a pastoral idyll; yet complex ecological relationships make constructing a natural cycle system challenging. Despite these difficulties, we still are developing methods and tools for learning from nature, requiring increasing numbers of researchers and practitioners to contribute in the future.

8.3 Design Cases Inspired by Nature

While the theoretical methods of learning from nature still need refinement, various design practices have become increasingly prevalent worldwide. These practices enable us to understand intuitively and clearly the various perspectives and rich possibilities of learning from nature design. Learning from nature's design involves not only imitating the appearance, color, and pattern of natural objects, but also understanding nature's mysteries from the perspectives of materials, forms, structures, principles, and systems. This section describes numerous cases of learning from nature, aiming to inspire further research and practice in this field (Fig. 8.7). These include (1): learning from the cycle of natural substances and the biological characteristics of materials; (2) imitating the form of natural objects; (3) learning nature's structure at both micro and macro levels; (4) grasping the operating mechanisms of natural things, from birth, adaptation, competition, cooperation, breeding to decline; and (5) learning from nature at the system level: the integration and interactions of material, form, structure, and principle features.

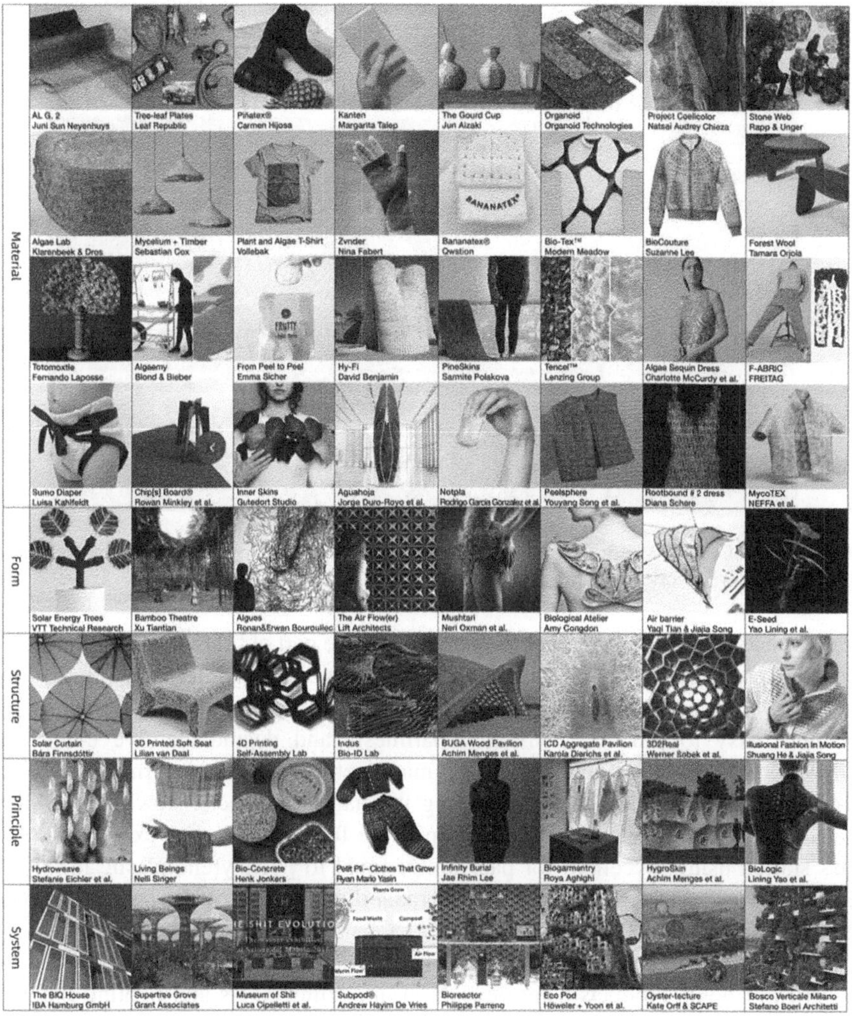

Fig. 8.7 Design cases entailing learning from nature

8.3.1 *Materials*

The Magic of Mycelium

In recent years, design application research on mycelium materials has launched on a global scale. Mycelium is the filamentous body of fungi that provides them nutrition, like the root hairs of mushrooms. It forms a tubular, filamentous, porous structure through symbiosis with a culture substrate (wood chips, straw, bran, grains, and so on), with lengths varying from a few microns to several meters depending on the

Fig. 8.8 EcoCradle®Mushroom™ packaging produced by Ecovative Company (Image source: https://matadornetwork.com, https://www.greenmatters.com)

fungus and culture environment.[42] As the vegetative part of fungi, mycelium adheres to substrates like an adhesive, maintaining its integrity even after death, thus forming new types of bio-composite material. The resulting materials are sturdy, lightweight, easy to process, non-toxic, harmless, and degradable. We could use mycelium not only as food, but also to manufacture packaging, furniture, houses, and even clothing. Designers leveraging the natural characteristics of mycelium introduce sustainable solutions while ensuring functionality. For example, IKEA has started using mycelium for packaging, equal to polystyrene plastic in impact resistance and compressibility; its biodegradability positions it as a potential substitute for plastic (Fig. 8.8).

British designers Sebastian Cox and Ninela Ivanova have designed and produced a series of home products using mycelium, including stools and lighting fixtures, branding their work Mycelium + Timber (Fig. 8.13). These two materials have a close relationship in nature. For thousands of years, mycelium has tightly wrapped the roots of trees, with trees relying on this vast mycelium network to transmit secret information unknown to humans. How do the designers make use of this relationship? They place discarded wood or sawdust into molds and add a fungus called Fomes fomentarius. The mycelium grows rampantly through the discarded wood, eventually assuming the shape of the mold.

Such applications of mycelium are not limited to indoor home products. We can also use mycelium to manufacture outdoor building materials. In 2014, designer

[42] Reference website: https://www.arts.ac.uk/subjects/textiles-and-materials/postgraduate/ma-biodesign-csm

David Benjamin collaborated with Ecovative Design to make 10,000 bricks from corn stalks and mycelium, and built the Hy-Fi mushroom tower pavilion, exhibited at the Museum of Modern Art in New York. The revolutionary aspect of this building is not its appearance, but Benjamin's bold attempt to introduce biomass materials into the field of architecture, indicating the prospects for commercial applications and potential value of this material.

Cellulose Fashion

Currently, the majority of raw materials for textiles rely on traditional plant materials or petrochemical products, which not only consume resources but also cause environmental damage. British designer Suzanne Lee, a prominent figure in sustainable fashion design, has proposed a novel approach to using natural products to create fashion. Her BioCouture design[43] uses cellulose cultured from bacteria, fungi, and other microorganisms for clothes from materials that are grown.

Suzanne Lee employs a kombucha recipe,[44] adding specially cultured bacteria and yeast to a sugary tea solution. Sugar is an indispensable ingredient, promoting the fermentation of biomass and providing energy for the bacteria. After 2–3 weeks, the cellulose filaments growing on the surface of the liquid adhere to each other, eventually forming a film about 1.5 cm thick. After drying and flattening, Lee processes the film using the same methods as with traditional textiles, cutting and sewing. The resulting material absorbs natural dyes well and has a texture similar to synthetic leather. At the end of its life, it can be composted along with vegetable waste. However, Lee's team is still researching ways to make the material waterproof and to control biodegradation to prevent premature rotting.

Using cellulose produced by microorganisms in fashion not only prevents resource waste, but it also makes efficient use of waste streams from the food and beverage industry. For instance, discarded sugar water from food processing plants serves as both nutrient and raw material. The resulting product eventually decomposes, returning to nature. This is a quintessential example of circular economy design. This does not mean that cellulose will replace cotton, leather, or other textiles. In terms of real applications, ecofriendly materials still have a long way to go. While these clothes may not become mainstream, it is conceivable that one day in the future, we may buy table lamps, chairs, or even houses grown by microorganisms.

Laboratory-Grown, Biologically Based

The global business value of leather used in clothing, footwear, luxury leather bags, and furniture manufacturing amounts to some $100 billion.[45] The unique texture, aesthetic appeal, and cultural symbolism of leather make it an important material, but

[43] Islam M R, Tudryn G, Bucinell R, et al. Morphology and mechanics of fungal mycelium [j]. *Research Support*, 2017, 7(1): 13–70.

[44] *Time* Magazine named this project one of the "50 Best Inventions of 2010."

[45] This is a symbiotic mixture of bacteria, yeast, and other microorganisms that produce cellulose during the fermentation process. Leather processing requires soaking, degreasing, depilation, ashing, deashing, softening, acid soaking, tanning, neutralization, retanning, dyeing, greasing, and so

also result in the death of billions of animals each year.[46] Obtaining leather through slaughter is not only cruel but also inefficient and has a significant environmental impact.[47] Modern Meadow, a biotech company based in New York, has cultivated leather in its laboratories using bio-manufacturing engineering and cell technology. Collagen, which has a texture, flexibility, and durability similar to those of traditional leather, is the core component of this laboratory leather. Laboratory-grown leather does not involve animal slaughter. Nor does it generate the various pollutants of the tanning process.[48] Protein-based Bio-Alloy™ and Bio-F@rm™ are the company's proprietary core technology platforms. Modern Meadow's leadership seeks to promote a global shift away from petrochemical and animal sources without compromising material performance. Bio-Tex™ is the first textile coating material system using Bio-Alloy™ technology, which, when coated on the back of the material, creates a tactile sensation akin to leather. It surpasses traditional tanned leather in color brightness and performance while reducing greenhouse gas emissions by over 90%.[49] Bio-Tex™ is a bio-based product certified by the US Department of Agriculture's BioPreferred® program.[50] Modern Meadow has begun collaborating with brands such as Tory Burch and Everlane and other bag brands to develop new products. We expect this bio-based leather to transition from the laboratory into our everyday lives shortly.

8.3.2 Morphology

Branching Units

Natural forms have fueled human imaginations and provided endless inspiration since ancient times. Following such inspiration, product designers Ronan and Erwan Bouroullec have imitated the form of branches and seaweed to design a home partition. By repeatedly adding and connecting branch units, they formed a soft partition that grows like a plant, and can expand, extend, and bend in various directions. The partition can also adjust its density to modulate the intensity of the light that permeates it through repetition and superposition. This design creates natural beauty and

on. The process is complex and the chemical materials used in the process extremely diverse, which has a significant impact on the environment.

[46] Tanning is an important process in leather and fur processing. Different tanning effects are produced using different tanning agents.

[47] Data source: https://www.modernmeadow.com/biomaterials

[48] The USDA BioPreferred® Program aims to help consumers easily identify commercially available bio-based products. The program was created under the 2002 Farm Bill. Its goal is to increase the purchase and use of bio-based products to promote economic development, create job opportunities, provide a market for agricultural products, and reduce America's dependence on oil.

[49] This research was the cover article in *NATURE* magazine in February 2023, LUO, D., Maheshwari, A., Danielescu, A. et al. Autonomous Self-Burying Seed Carriers For Aerial Seeding. Nature 614, 463–470 (2023). https://doi.org/10.1038/s41586-022-05656-3

[50] Qiu Song. *Research and Application of Design Morphology*[M]. Beijing: China Architecture & Building Press, 2019: 10–18.

not only has the functionality of freely adapting to any space, but also has a form that resonates emotionally with users.

Artificial Super E-SEED

The Storksbill (Erodium) plant has seeds shaped like drill bits. When it rains or there is an increase in humidity, the tightly coiled bodies of the seeds automatically loosen, twisting the seeds into the soil, thus avoiding predation by birds and other animals, and improving their growth environment. Inspired by this natural phenomenon, Carnegie Mellon University, Syracuse University, the University of Pennsylvania, Zhejiang University, Accenture Labs, and several other institutions collaborated to design and develop a biodegradable self-drilling seed carrier, E-seed, to increase the germination rate of aerial seeding. Similar in shape to the Storksbill, the E-seed has a body tightly wound like a drill bit and a coiled tail. When exposed to moisture, E-seeds expand and rotate, burying their heads in the soil, optimal for those of around 11 millimeters in length and about 72 milligrams in weight, such as vegetable and vegetation seeds. Under flat soil conditions, the success rate of seed drilling can reach 80%.[51] To increase the applicability and efficiency of E-seeds in natural environments, the research team also developed a three-tailed E-seed. Applying of this design may aid in preventing landslides, reducing species invasion, and aiding vegetation restoration. At the same time, the team can also implant E-seeds with sensors to monitor the environment and assist exploration of nature (Fig. 8.9).

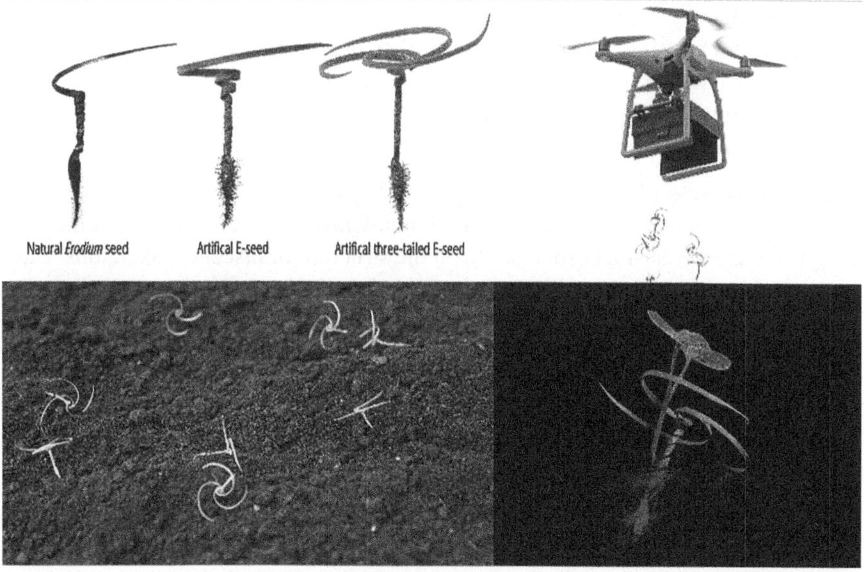

Fig. 8.9 Storksbill and the E-SEED (Image source: https://design.zju.edu.cn/guanyunlab/)

---

[51] Kapsali V. *Biomimetics For Designers: Applying Nature's Processes & Materials In The Real World*[M]. London: Thames & Hudson Ltd., 2016:214–219.

Fig. 8.10 Research on insect exoskeleton for crash-resistant drone design (Image source: Kang Pengfei)

8.3.3 Structure

3D-Printed Soft Chair

Dutch designer Lilian van Daal mimicked the structure of plant cells and created a unique chair using 3D printing. Van Daal studied the structural methods of various plant cells so that she could ensure the chair's elasticity and comfort by altering the structure's density without using cushions or fillers. Due to her clever structural design and 3D printing process, the chair only contains one type of raw material and is super light, which maximizes savings in material costs and prevents waste generation during manufacturing, reduces emissions and pollution during transportation, and facilitates product recycling and remanufacturing.

Insect Exoskeleton Drone Crash-Resistant Design

Kang Pengfei, a graduate student at Tsinghua University's Academy of Fine Arts, designed a crash-resistant drone based on the structure of insect exoskeletons. Kang conducted in-depth observations, research, and experiments on the exoskeletons of Coleoptera, discovering that the structure of their shells effectively resists pressure and impact. He then summarized the morphological rules and mechanical structure characteristics of the insect and applied them to the main body structure design of a drone, greatly enhancing the aircraft's crash resistance in both longitudinal and vertical impacts (Fig. 8.10).[52]

[52] In the cool mornings in the desert, their skin is usually dark to promote heat absorption from the sun; as the temperature and their body temperatures rise during the day, their skin begins to turn white or light gray to reflect sunlight. B. Michael Walton, Albert F. Bennett. Temperature-Dependent Color Change In Kenyan Chameleons.[J]. Physiological Zoology, Vol. 66, No.2, 1993: 270–287.

8.3.4 Principle

Climate-Sensitive Pavilion

The process and principles of nature also inspire us. Based on the functional characteristics of spruce cones, Professor Achim Menges at Germany's University of Stuttgart created an architectural skin that can respond to climate changes and built a small Climate-Sensitive Pavilion (Hygroskin: Meteorosensitive Pavilion). Spruce trees belong to the pine family. Their cones change shape according to humidity level, opening when the air is dry and closing when it is humid. Imitating them, the pavilion resembles a living building, where each unit can automatically change shape according to current climate conditions, making it a dynamic structure in which the openness and porosity of the units constantly adjust, thereby regulating light transmission and visual permeability.[53] The architectural skin uses the humidity-sensitive characteristics of its plywood composite to respond to weather changes, without the need to provide operating energy or any mechanical or electronic control. Unlike high-tech devices, this building follows a brilliant no-technology strategy, perfectly integrating material, form, and function.

Temperature-Adjusting Clothing

The common method for regulating human body heat is to take off or put on clothing. Meanwhile, animals and plants in nature use other mechanisms to regulate their body temperatures. Among these is a desert-dwelling chameleon (Namibian chameleon)[54] the skin color of which changes not to communication or a form of camouflage, but to regulate body temperature. In 2022, an interdisciplinary team from Tsinghua University designed Thermal Responsive Clothing that mimics the Namibian chameleon by using thermodynamic ink[55] to create clothes that self-regulate according to the surrounding temperature (Fig. 8.11). Temperature changes activate the thermodynamic ink. When the temperature rises, the ink lightens, revealing the graphic. The light shade reflects heat and cools the wearer. When the temperature drops, the ink turns black and begins absorbing heat and warming up. Currently, the temperature control effect of the fabric coated with this ink is ±5 °C, with a maximum temperature difference of up to 11 °C. The ink reacts to two temperatures, 15 °C and 22 °C, to enhance the adaptability of temperature control in different seasons and outdoor environments. This project explores autonomous temperature control in outdoor clothing through integration of materials, technology,

---

[53] Thermodynamic ink is composed of heat-responsive, scattering nanocrystals and graphene layers. It can effectively reflect environmental light sources and radiate its own infrared heat to achieve cooling at high temperatures. In a low-temperature environment, the ink presents a black surface with a strong light absorption and heat preservation, thereby achieving comfortable constant temperature control in high- and low-temperature environments.

[54] Volstad N L, Boks C. On The Use Of Biomimicry As A Useful Tool For The Industrial Designer [J]. *Sustainable Development*, 2012, 20(3): 189–199.

[55] Ceschin F, Gaziulusoy I. *Design For Sustainability: A Multi-Level Framework From Products To Socio-Technical Systems* [M]. London: Routledge, 2020.

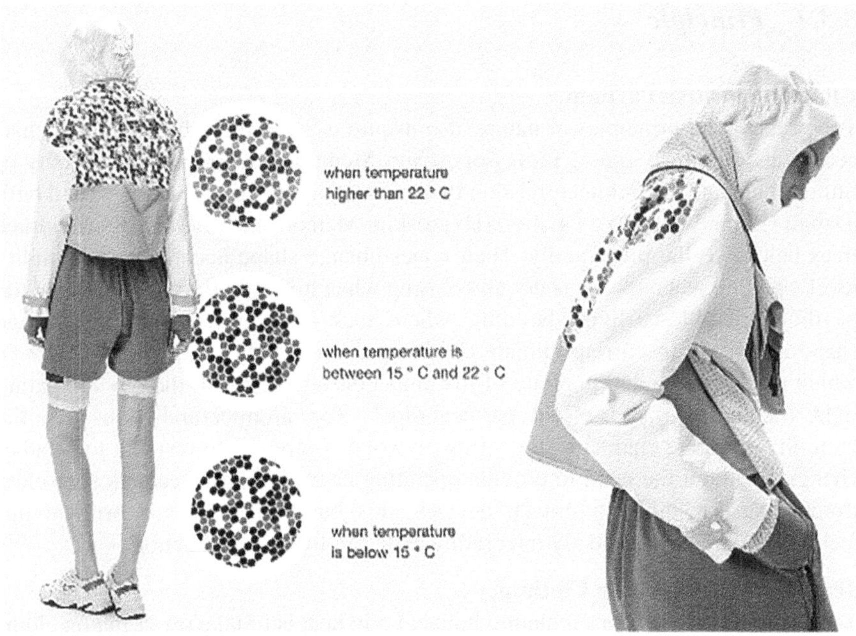

Fig. 8.11 Thermal control ink coating and thermal responsive clothing (Image source: https://www.ad.tsinghua.edu.cn/info/1061/28697.htm)

and design, meeting functional needs, improving efficiency, and reducing clothing waste and resource consumption. The team adopted zero waste and detachable design strategies in the design and cutting processes to make the clothing sustainable.

8.3.5 Systems

Supertree Grove

The world's ecosystem comprises a series of complex and subtle relationships, which humans can only imitate to some extent. The Gardens by the Bay in Singapore is a large architectural and landscape community, which features the Supertree Grove designed by Grant Associates (Fig. 8.12). This design fully embodies interdependent, cooperative relationships in natural systems. The Supertree Grove consists of 18 supertrees, each 25–50 meters high, with four main parts: a reinforced concrete base, a trunk, a planting board, and a canopy. Various plants, tropical climbing plants, epiphytes, and ferns, are planted on the trunk. The design of the Supertree Grove combines the shape and function of trees, not only supporting plant growth but also mimicking the ecological function of real trees.

Fig. 8.12 Supertree Cluster (Image source: https://en.wikipedia.org/wiki/Gardens_by_the_Bay)

During the design phase, each supertree received a different function, creating a simulated forest. For example, the canopy is equipped with photovoltaic cells to absorb solar energy for nighttime lighting. Some canopies serve as exhaust ports connected to the plant cooling system, while others act as intake ports. The entire Supertree Grove system drives airflow, creating a breeze at ground level (Fig. 8.13). In summer, the canopies provide shade and ventilation for people, while also regulating the ground temperature. The designers aimed to create a world similar to a real natural forest ecosystem, while minimizing the consumption of resources.

8.4 Inspiration from Hidden Knowledge

Various design practices that entail learning from nature have infinite potential and provide new inspirations for future human development. These practices not only have significant value for sustainable design, but also yield new ideas. For example, development and application of biodegradable materials can reduce the consumption of non-renewable resources and waste (especially plastic waste), thereby mitigating the negative impact of human activities on the ecotope and potentially providing new fashion and business opportunities. Drawing on natural forms to create adaptive, flexible design units can reduce the cost of product manufacturing, transportation, and installation, improve efficiency, and reduce energy

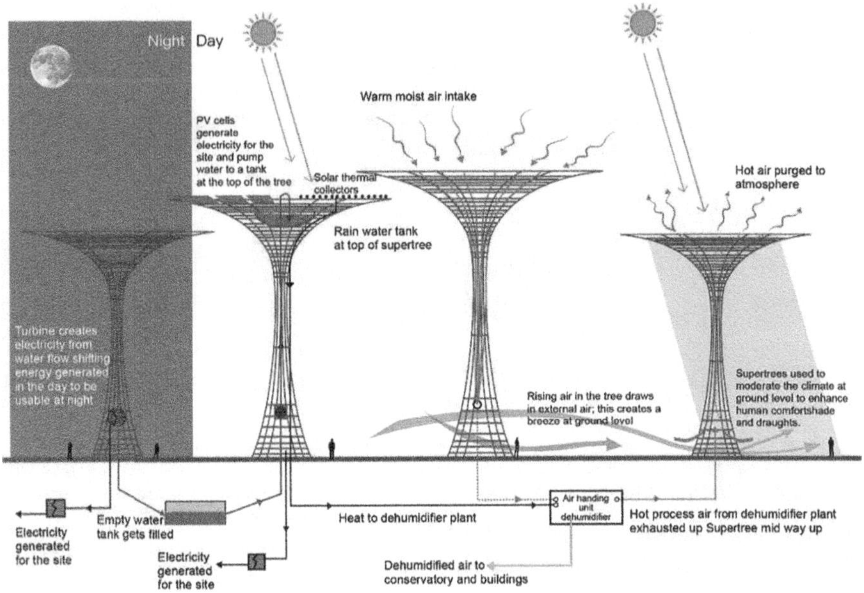

Fig. 8.13 Operation of Supertree Cluster (Drawn based on image provided by Grant Associates Studio)

consumption, while also introducing natural beauty and emotional comfort to people in the digital age. Learning from nature's structures can create robust, durable, lightweight, efficient products that meet functional goals, while reducing consumption of materials, thereby lowering manufacturing and recycling costs and creating unique product value and aesthetic experiences.

Understanding and replicating the operating mechanisms and principles of nature can lead to creation of adaptive, self-regulating, self-empowering natural objects. This requires understanding the dynamic life processes of natural entities, from birth, adaptation, competition, cooperation, and breeding to decline, and harnessing life wisdom for the highest ecological efficiency. System-level learning from nature calls attention to the connections and interdependencies among different natural entities, focusing on integrating materials, forms, structures, and principles. Natural systems are perfect closed-loop systems, epitomizing the Cradle-to-Cradle (C2C) concept and circular economy model. System-level design emphasizes the balance and alignment of material, energy, and information flows among elements of a system, aiming for self-sufficiency that provides nourishment to other links in the system. System-level design represents the highest form of learning from nature and also the most challenging.

Although humans have effectively drawn on nature for inspiration since the beginning of our existence, claiming that innovations inspired by nature are inherently sustainable is misleading.[56,58] Most of the abovementioned design practices are still in the experimental or prototype testing stage. Few products have reached the market. We still lack comprehensive evaluations of the environmental friendliness, economy, and utility of specific products. Additionally, strategies and methods we derive from nature primarily serve as concept generators that inspire design, making it difficult to accurately measure and evaluate results. For example, the widely recognized idea that cities should operate like ecosystems is one inspired by nature. But how can we assess the sustainability of such an ecosystem? Although we can assess the ecological benefits of a system using tools such as lifecycle assessment (LCA), the complexity of cities and numerous related parameters (especially social and cultural factors) make it challenging to accurately evaluate sustainability.

Furthermore, learning from nature is both a strategy and an interdisciplinary activity. Design creates connections, but is constrained by the extent and depth of interdisciplinary cooperation. For instance, designers without training in biology may struggle to find biological solutions relevant to design problems.[59] Although databases exist that contain information on biological designs, understanding the principles and mechanisms of biology is not easy. However, with advances in the life sciences and biological control research, as well as the dissemination of knowledge and technology, designers will have increased access key information, potentially broadening the range of their potential interests and creativity.

As stated at the beginning of this book, there are no purely sustainable solutions. The mission of designer is to perceive problems and have the courage and wisdom to explore new opportunities. Such explorations will inevitably involve failures, but this does not diminish the value of the explorations. The strategy of learning from nature opens a door for designers. Delving into this strategy reveals the vast, boundless territory of natural knowledge, far beyond our cognitive range and ability. This entails a long process of constantly searching for treasures. Even when we can perceive them, it is not easy to come by nature's gifts. Often, nature's inspiration is subtle, requiring keen perception, a wise mind, and diligent hands to discern and explore. If all future human innovations (technological, economic, and social) consciously drew upon nature's mechanisms and laws, actively establishing an inherent connection with nature, then we would have reason to anticipate a more future.

---

[56] Buck N T. The Art Of Imitating Life: The Potential Contribution Of Biomimicry In Shaping The Future Of Our Cities [J]. *Environment And Planning B: Urban Analytics And City Science*, 2017, 44(1): 120–140.

Chapter 9
Product-Service System Design

Abstract The Product-Service System (PSS) Design is an innovative strategy that shifts from merely designing and selling material products to providing system solutions that encompass both products and services. A well-designed PSS can effectively reduce resource consumption and corporate costs, while simultaneously creating new opportunities for profit growth and improving responses to consumer needs.

Keywords Service · Product-service system · Sharing economy · Dematerialization

As discussed in Chap. 2, solely relying on changing the design of physical products cannot fully address environmental issues. This is due to the insufficient intrinsic motivation among enterprises to promote green and ecological design, which may lead to intentional or unintentional greenwashing. Furthermore, even if companies focus on designing and manufacturing green products, continued adherence to the economic model of mass production, mass consumption, and mass disposal will result in significant negative impacts to the environment. The innovative business models and consumption patterns inherent in PSS design, such as supplier coordination mechanisms and changes to product leasing and sharing models, offer a systematic restructuring of production and consumption. This restructuring can significantly reduce the environmental impact of economic development, promoting transition to a sustainable society.

9.1 Origin and Definition

9.1.1 Service Design and Product-Service System Design

Since the mid-twentieth century, the world's economic structure has undergone a significant transition. The manufacturing industry's dominance has waned in Western countries, giving way to the flourishing of various emerging service

X. Liu et al., *Design for Sustainability*, Sustainability and Interdisciplinarity,
https://doi.org/10.1007/978-981-96-8657-5_9

sectors. In 1968, American economist Victor R. Fuchs introduced the concept of the service economy in his book *The Service Economy*. During this period, modern society gradually adopted a post-modern form. The development of information technology and the internet led to changes in public consumption patterns, steering an industrial society toward dematerialization and a focus on services.

In the 1980s, G. Lynn Shostack first proposed designing services from the perspective of marketing and management.[1] Gillian and Bill Hollins introduced the concept of service design in their 1981 book, *Total Design: Managing the Design Process in the Service Sector*. That same year, Michael Erlhoff and Birgit Mager of the Cologne International School of Design (KISD) began integrating service design into design education, dedicating themselves to related teaching practices and research.[2]

By the 1990s, driven by the popularization of green environmental concepts and public pressure, enterprises and designers started integrating environmental design requirements into product development. However, the design community gradually realized that without changing the architecture of the traditional production and consumption system, the strategy of optimizing product lifecycle design alone could not fundamentally reduce environmental impact. Consequently, the focus of the design community has shifted from the environmental efficiency of products to system innovation with ecological benefits. Design researchers began exploring concepts related to immaterial design and developing new design rules and methods. In 1993, Denmark hosted the Mind over Matter: Immaterial Design in the Age of Material Limits International Industrial Design Conference, during which participants expanded the boundaries of design from traditional industrial products to new fields such as digitization, information, and services. In his 1992 book, *The Immaterial Society: Design, Culture, and Technology in the Postmodern World*, Marco Diani notes that technological changes and the social and cultural changes that accompany them reflect a "gradual transition from a society based on manufacturing material products to a society based on producing services or immaterial products."[3] This context gave rise to the proposal of PSS design as a strategy for immaterial innovation.,[45]

---

[1] Shostack, G. L. How To Design A Service[J]. *European Journal Of Marketing*, 1982, 16(1): 49–63.

[2] Xin Xiangyang, Cao Jianzhong. Positioning Service Design[J]. *Packaging Engineering*, 2018, 39(18)43–49.

[3] Diani, M. *The Immaterial Society: Design, Culture, And Technology In The Postmodern World*[M]. Englewood Cliffs, NJ: Prentice Hall, 1992: 6

"transformations from industrial to postindustrial societies, reflecting changes from a society based on manufacturing and producing material goods to a society based on a service economy (immaterial goods)."

[4] Vezzoli C. *System Design For Sustainability: Theory, Methods And Tools For A Sustainable "Satisfaction System" Design*[M]. Rimini, Italy: Maggioli Editore, 2010.

[5] Vezzoli C, Kohtala C, Srinivasan A, et al. *Product-Service System Design For Sustainability* [M]. Sheffield, UK: Greenleaf Publishing, 2014.

In the twenty-first century, the concept of and methods for implementing PSS have expanded beyond academic research with applications to transforming business models. Facing market contraction and intensified competition, traditional manufacturing executives began viewing services as a new avenue for profit growth. They realized that combining products and services can yield higher profits than selling products or services separately, fundamentally changing their perspective on marketing. With the convergence of the servitization of products and the productization of services, PSS design has gained increasing importance, leading to the emergence of influential projects and industries.

9.1.2 Definition of Product-Service Systems

The United Nations Environment Program (UNEP) defines a product-service system as the result of a *shift from simply designing and selling material products to providing a comprehensive product and service system to better meet people's specific needs.*[6] Because of its characteristics as a win-win situation incorporating both economic development and environmental benefits, PSS is a concept with great potential for sustainable transformation to an ecological model with high efficiency. Its core idea is *enterprises provide functions or product results, allowing users to enjoy value without physically owning products.*

Since the proposal of PSS, various scholars have interpreted it differently based on their own priorities. In the 1990s, scholars saw PSS simply as a supplement to businesses, a system composed of products, services, platforms, and maintenance equipment aimed at improving the ability to meet user needs and enhancing corporate competitiveness. By the early twenty-first century, PSS had evolved into a corporate strategy, shifting from selling physical products to providing comprehensive products and services that meet specific customer needs.[7] Table 9.1 presents the various definitions of PSS:

In summary, the best definition of a PSS is as *an integrated system of products and services one or more socioeconomic participants provides that aims to meet a specific customer demand.* The term system refers both to the product and service

---

[6] UNEP. *Product-Service Systems and Sustainability: Opportunities for Sustainable Solutions* [M/OL]. United Nations Environment Program, 2002: 8. https://wedocs.unep.org/handle/20.500.11822/8123.

"This PSS strategy offers a customised mix of services (as a substitute for the purchase and use of products), in order to provide a specific final result (in other words, an integrated solution to meet the customer's satisfactions)."

[7] Manzini E, Vezzoli C. A Strategic Design Approach To Develop Sustainable Product Service Systems: Examples Taken From The "Environmentally Friendly Innovation" Italian Prize[J]. *Journal Of Cleaner Production*, 2003, 11(8): 851–857.

"A product service system (PSS) can be defined1 as 'an innovation strategy, shifting the business focus from designing (and selling) physical products only, to designing (and selling) a system of products and services which are jointly capable of fulfilling specific client demands'"

Table 9.1 Definitions of Product-Service System (Vezzoli C, Kohtala C, Srinivasan A, et al. *Product-service system design for sustainability* [m]. Sheffield, UK: Greenleaf Publishing, 2014: 30)

Authors	Year	Definition
Goedkoop, van Halen, te Riele, Rommens	1999	A product-service system (or combination of products and services) is a set of marketable products and services jointly capable of fulfilling a need for a client. The PSS may lead to a benefit for the environment in connection with the creation of a (new) business.
Mont	2002	PSS is a system of products, services, networks of actors and supporting infrastructure that continuously seeks to be competitive, satisfy customer needs, and have a lower impact than traditional business models.
UNEP: Manzini, Vezzoli	2002	Result of an innovative strategy that shifts the center of the business design and sale of products only to systems offering products and services that are jointly capable of satisfying a given application.
Brandstotter	2003	PSS is a product of material and intangible services designed and combined so that both jointly are able to satisfy a specific need of a user. In addition, a PSS may reach sustainability targets.
EU, MEPPS: Van Halen et al.	2005	Result of an innovation strategy focused on the design and sale of a system of products and services that are jointly capable of fulfilling a specific customer demand.
Baines et al.	2007	PSS is an integrated offering of a product and a service that provides a value. Using a PSS offers the opportunity to decouple economic success from material consumption and thus reduce the environmental impact of economic activity.
UNEP: Tischner, Vezzoli	2009	System of products and services (and infrastructure), to jointly cope with the needs and demands of customers in a more efficient way with better value for both businesses and customers, compared to only offering products....PSS can decouple the creation of value from the consumption of materials and energy and thus significantly reduce the environmental impact in the lifecycle of traditional product systems.

system the customer receives and to the participant system that produces or provides the combination of products and services. Therefore, a PSS is a specific value proposition that emphasizes meeting customer needs through delivering functionality rather than selling products or services.

Generally speaking, a PSS includes five key elements:[8,9,10,11]

[8] Mont O. Clarifying The Concept Of Product-Service System[J]. *Journal Of Cleaner Production*, 2002, 10(3): 237–245.

[9] Muller P, Stark R. Detecting and structuring requirements for the development of product- service systems[j]. symposium "design for x" Neukirchen, 09–10 OCT., 2008.

[10] Ceschin, F. *Sustainable product-service systems: Between strategic design and transition studies*[m]. Springer Science & Business Media, 2013: 21.

[11] Yu Le, Li Binbin. Research on Product Service Design from the Perspective of Sustainability[J]. *Packaging Engineering*, 2011, 32(20): 73–76.

1. *Products:* Tangible artifacts in the system (automobiles, washing machines, electric meters, etc.);
2. *Service:* Various services within a product's lifecycle (sales, leasing, sharing, maintenance, upgrades, recovery, etc.);
3. *Collaborator (Participant) Network:* All socioeconomic participants in the system and their relationships to each other (producers, service providers, energy providers, managers, users, etc.);
4. *Policies and Contracts:* Charging strategies, management norms, and behavioral standards between business collaborators and between collaborators and consumers regarding product use, leasing, and recovery;
5. *Infrastructure:* Public and private facility systems (roads, communication lines, waste collection systems, etc.).

Implementing a PSS means simultaneously designing both these elements and the relationships between them. Designers must shift from product design thinking to system design thinking.

9.2 Features of Product-Service Systems

9.2.1 From Lifecycle to Ecological Benefits

In traditional production and consumption systems, innovations in material products are the core means for enterprises to compete and develop. As we have repeatedly noted, relying solely on such innovations is unsustainable for both business and society as a whole. Manufacturing executives and suppliers cannot truly care about the energy consumption of the people who use their products. Nor are they interested in the environmental impact of and resource waste discarded products cause. Indeed, manufacturers often sell products with short lifecycles on purpose to increase profits. Although many companies have implemented lifecycle optimization design, applying various green and ecological design strategies, and strive to minimize resource consumption and environmental impact, the pursuit of maximizing their own economic gain prevents them from realizing ecological benefits to the overall system. Although product energy efficiency has increased, the increase in usage has resulted in an overall increase in energy consumption. The comparison data mentioned above about the simultaneous increase in both automobile energy efficiency and mileage is a case in point.

Taking laundry as an example. Doing laundry involves multiple, related objects: washing machines (products), detergents (materials), maintenance (services), and electricity (energy consumption). If viewed from the perspective of a single product, strategies to reduce resource consumption and environmental impact must rely on independent modules, such as production, sales, transportation, and recycling. Product lifecycle design can only provide the opportunity for product designers, such as washing machine and detergent manufacturers, to optimize at their own level. However, based on a pay-per-use model, laundry service as a system solution allows users to enjoy the use of washing machines, machine maintenance, and upgrade services without having to purchase a washing machine, providing a complete service opportunity and meeting more needs than purchasing products does. This satisfies the needs of enterprises, consumers, and the environment all at once, generating a system innovation based on ecological benefits through restructuring, balancing, and benefits to different stakeholders, overall reducing the negative impact on ecology.

By establishing a production and consumption system based on interactions between and cooperation among different participants, sustainable PSS design has the potential to minimize environmental impact during a product's lifecycle. This interactions between stakeholders can promote resource optimization with the entire system meeting specific needs and achieving ecofriendliness. Innovations occur not only at the product level, but also in terms of the relationships among various stakeholders, creating new socioeconomic cooperation networks.

Based on resource integration optimization of entire systems, no longer that of a single link, PSS design seeks to meet user needs by integrating products and services. Rational, efficient system design will not only help enterprises effectively obtain economic value, but also effectively reduce product purchases and lower resource consumption, improving our use of resources[12] and benefiting the environment.

9.2.2 *System Innovation Based on Functional Economy*

PSS replaces products with services (although the degree of replacement varies), with attention on users' needs, rather than selling a certain type of product. This model allows for full exploitation of the use value of a product, thereby achieving the goal of efficient use of resources. Another tangible model replaces consumers' needs, creating new systemic innovations based on value acquisition and service provision.

When users' needs are no longer just a product, but rather a function, such as clean clothes, a comfortable temperature, convenient travel, or moderate lighting, then washing machines, air conditioners, automobiles, and lamps are no longer the sole objectives and other elements in the system and their supply and configuration methods become the focus of attention. Some scholars believe that the PSS belongs to the functional economy, the foundation of which is to meet consumer needs by providing system functions rather than functional products. In a functional

[12] Tukker, A. Eight types of product–service system: Eight ways to sustainability? Experiences from suspronet[J]. *Business Strategy And The Environment*, 2004, 13(4): 246–260. "Figure 1. Main and subcategories of PSS"

economy, manufacturers and suppliers charge based on satisfaction with the delivered service, not the units of products sold. Therefore, a functional economic system can also promote economic growth by reducing and optimizing material and energy consumption. This approach allows meeting human needs and promoting economic development to completely decouple from resource consumption and environmental pollution.

9.3 Classification of Product-Service Systems

Enterprises providing pure products (such as automobiles) or pure services (such as transportation by bus) have a long history, and combining products and services is nothing new (such as post-sale warranties for automobiles). However, in recent years, due to different goals and value orientations, the forms in which products and services combine have become increasingly varied. Classifying these by type thus helps deepen our understandings of the characteristics and sustainable potential of PSS. In 2002, a UNEP report classified three types of sustainable PSS: product-oriented (services that provide added value to extend a product's lifecycle), use-oriented (services that provide a platform for users), and result-oriented (services that generate results for users). These three types of PSS are located between physical products and service experiences, revealing various states and modes of transition from tangible value (product) to intangible value (service). Scholars such as designers Arnold Tukker and Ursula Tischner provide corresponding classification models, as shown in Fig. 9.1.

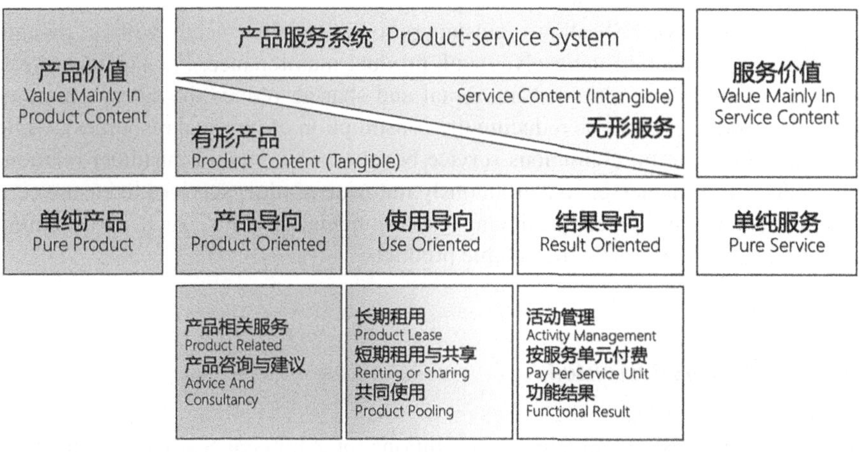

Fig. 9.1 Classification of Product-Service Systems (Vezzoli C, Garcia Parra B, Kohtala C. *Designing Sustainability For All: The Design Of Sustainable Product-Service Systems Applied To Distributed Economies*[M]. Springer Nature, 2021: 6)

9.3.1 Product-Oriented

Product-oriented PSS provide additional services to users to ensure the performance of products throughout their lifecycles, including repair, replacement of parts, upgrades, replacements, recycling, and various post-sale services for products. The product is still at the core of this model, with services providing it added value. The primary forms of this type of PSS include product-related services, advice, and consultancy services.

This is the typical product-service method. Its purpose is to realize the function of a product, while improving efficiency of use, reducing consumers' burden and responsibility when learning to use a product, as well as recycling it, thereby increasing consumers' trust in a brand and their brand loyalty. Because it expands the business scope of product suppliers, it may drive corporations to seek environmentally beneficial and increasingly competitive business solutions. Economic benefits result not only from product sales, but also from comprehensive, thoughtful services, increasing the stickiness of the product and the value of the brand.

9.3.2 Use-Oriented

Use-oriented PSS provide a service platform (products, tools, opportunities, or capabilities) that meet people's specific needs and desires. Products and services coexist, with products effectively integrated into the service. Consumers use but do not need to own the product, only paying for an agreed-upon time period or costs of use, such as automobile rentals, shared bicycles, self-service laundromats, etc. The main forms of use include long-term product leasing, short-term product rental/sharing, and joint product use (pooling).

This category of PSS allows consumers to satisfy their needs without owning products or concerning themselves with product maintenance. For manufacturers, operators, and other value owners, rental and sharing reduce the sales volume of physical products (likewise reducing the consumption of physical resources), while generating long-term, continuous service benefits. This type of customer relationship may drive businesses to continuously innovate product services to ensure economic benefits, ultimately achieving environmental benefits, such as designing efficient, durable, reusable, recyclable products.

9.3.3 Results-Oriented

Result-oriented PSS entail integrated solutions for achieving customer satisfaction. Value providers (company, supplier, stakeholder group, etc.) deliver customized, comprehensive solutions (as an alternative to buying and using products) to meet customers' special needs, such as providing efficient public transportation, outsourcing cleaning services, offering suitable temperature and lighting services for buildings, etc. The main forms of solutions include pay-per-service unit, activity management services, and functional result services.

In this configuration, the service becomes the core offering. Customers do not need to buy associated products, worry about maintenance, or even operate products themselves. They only pay according to whether they are satisfied after using the service. The enterprise retains ownership of products, usually provided by a third-party operating company to provide a specific final result (such as laundry service), rather than selling, renting, sharing products (such as washing machines). In other words, customers and manufacturers, suppliers, and operators agree on the results and cost of service, without a priori indicating which product will generate those results. According to this model, customers only need to focus on their satisfaction with the outcomes; companies owning the product will inevitably strive to improve resource utilization and productivity to achieve economic benefits and competitive advantages, such as designing and manufacturing more durable, reusable, recyclable products, which also generates environmental benefits.[13]

9.4 Case Studies of Product Service Systems

Already integrated into multiple industries, PSS design strategies have produced numerous excellent applications. Below, we will analyze cases belonging to the three categories of PSS (product-oriented, use-oriented, and result-oriented) in order to help readers deeply understand the sustainable value of PSS design.

Huawei's Upgrade Service (Product-Oriented)

Huawei is a consumer terminal company that excels at cutting-edge wireless communication and intelligent technology. In recent years, it has launched a series of hardware replacement, upgrade, and maintenance services aimed at retaining old users and enhancing user experience that cover almost all of the mainstream products it sells. Customers can receive personalized services such as replacement of original parts and device expansion at relatively low prices (mostly at cost), including glass panels, body memory, and built-in batteries, as well as free cleaning, film application, and other amenities. The service extends the lifespan of Huawei's products, not only giving old devices a new lease on life, avoiding the economic cost of replacing new phones, but also meeting users' demands for increasing quantities of storage for their pictures, audio and video data. Beginning in 2021, Huawei extended these complementary services from its mobile phone product field to other Huawei HarmonyOS products, including notebooks, tablets, and smartwatches.[14] The ser-

---

[13] Huawei Official Website: https://consumer.huawei.com/cn/support/

[14] Zhang L, Zhang J, Duan Z, et al. Sustainable Bike-Sharing Systems: Characteristics And Commonalities Across Cases In Urban China[J]. *Journal Of Cleaner Production*, 2015, 97:

vices have garnered positive reception and praise from old users for enhancing their user experiences and consequently consolidating Huawei's market share. Huawei's offerings represent a typical product-oriented hardware and software upgrade service.

Shared Bicycles (Use-Oriented)

With the rise and development of the sharing economy, shared bicycles have emerged as a green travel method. This time-sharing rental model has garnered widespread public acceptance. Shared bicycles offer users an economical, convenient solution for short-distance travel, obviating the need to own one's own bicycle. Instead, users can access bicycles when needed by finding and unlocking them through a mobile phone information management platform.[15] After completing their travel, users can park the bicycles in designated locations and automatically pay via an app. This model exemplifies a use-oriented product-service system. When designed well, managed properly, and operated in an orderly fashion, shared bicycles can save material resources, promote green transportation, reduce environmental impact, and meet people's short-distance travel needs. Additionally, they generate commercial benefits for enterprises.

Shared bicycles in the Chinese market have undergone four stages of development to date[16]:

First Stage (2007–2010): Using city administrations to manage the system, the Chinese government began introducing foreign public bicycle models domestically, mostly using bicycles that require docking stations.

Second Stage (2010–2014): Specialized bicycle market companies began to appear, although the public bicycles were still primarily docked.

Third Stage (2014–2018): With the rapid development of the mobile internet, internet-shared bicycles, led by Ofo and Mobike, emerged. Bicycles that did not require docking began to replace docked bicycles, increasing the convenience of the service.

Fourth Stage (post-2019): Internet brands such as Alipay, Meituan, and Didi entered the field. Early shared bicycle companies either transformed or underwent acquisition by large internet companies. Shared transportation forms diversified as electric bicycles appeared on the market, satisfying consumers' diversified travel needs.

In the past decade, internet-and-mobile-payment-supported shared bicycles have experienced capital chasing, explosive growth, frequent problems, public condemnation, and processes of transformation and reorganization. Initially, multiple brands competed for a share of the market, resulting in the convenience of copious choices for users. However, capital-driven disorderly competition led to excessive deployment of bicycles in specific geographical areas. Along with this, there was a

124–133.

[15] https://www.chinadaily.com.cn/business/2017-06/14/content_29731710.htm

[16] NIO Official Website: https://www.nio.cn

severe lack of systematic management, product maintenance, updates, and post-processing. Riders' ignorance of the rules of use and indifference to objects they do not own resulted in bicycles abandoned on a large scale, creating bicycle graveyards and causing traffic congestion and disorder. With increased government management efforts and enterprise transformation, some brands have begun to invest in environmentally friendly actions throughout the products' lifespan, implementing the 3R principle (Reduce, Reuse, Recycle) in all aspects of design, procurement, production, deployment, operation, and scrapping. The new generation of shared bicycles has also strengthened electronic parking management and optimized charging models, rekindling residents' enthusiasm in using shared bicycles.

NIO's BaaS Service Plan (Use-Oriented)

NIO, a Chinese automobile manufacturer, officially launched the BaaS (Battery as a Service) battery rental service in 2020. This business model, centered around battery rental, has created a PSS for battery charging and swapping, on-demand battery rental, and power upgrades, combined with NIO's offline service network system, to meet users' needs for batteries and flexible upgrades.[17]

While the concept has long existed, NIO has implemented it in real-life scenarios. The BaaS service addresses consumers' concerns when purchasing electric vehicles, targeting pain points such as high battery costs, anxiety about curtailed ranges of travel, and rapid battery performance decline and depreciation. The sales strategy combines automobile purchases with battery leasing, allowing consumers to rent rather than purchase batteries. BaaS users pay monthly rent for batteries, significantly saving on the initial costs of purchasing a vehicle, and can upgrade to a larger capacity battery pack at any time, which allows them to increase their range and meet long-distance travel needs.[18] NIO's BaaS strategy effectively lowers the threshold for automobile purchases and provides space for future upgrades. On-demand purchase, leasing, and flexible upgrades constitute NIO's energy service system.

As the core component of electric vehicles, the battery is costly and can pose risks to consumers if improperly maintained. NIO's BaaS service achieves separation of vehicle from battery, allowing consumers to lower the cost of purchasing new energy vehicles without worrying about battery degradation and upgrades. NIO thus increases product sales, extends its economic returns, strengthens its brand image, and improves user loyalty. Since it owns the battery, NIO is responsible for upkeep and disposal, ensuring the product is durable, easy to maintain, and its materials are recycled, reducing consumption and pollution emissions. Green environmental protection measures dictate achieving product closed-loop recycling within

---

[17] https://www.nio.cn/nio-power

[18] UNEP. *Product-Service Systems And Sustainability: Opportunities For Sustainable Solutions* [M/OL]. United Nations Environment Program, 2002: 14. https://wedocs.unep.org/handle/20.500.11822/8123.

"In developing countries this may have positive implications, by enabling greater involvement and ownership of local economies by local rather than global stakeholders."

the company, thereby minimizing environmental impact. NIO's BaaS is a typical use-oriented PSS design with significant sustainability potential.

Nigeria's Solar Home Systems (Result-Oriented)
The United Nations' seventh Sustainable Development Goals (SDG) is to ensure access to affordable, reliable, sustainable modern energy for all. As the most populous country in Africa, Nigeria's electricity demand is increasing, but the lack of power infrastructure and limited financial allocations for upgrades and updates has created a power supply bottleneck. Small-scale, distributed renewable energy is becoming the preferred local power solution. Although environmentally sustainable, renewable energy power equipment is unaffordable for ordinary families. Applying a PSS design strategy converts the cost of purchasing equipment into the cost of purchasing services, spreading that cost over an extended period of time, effectively helping local residents solve practical problems.

In December 2020, Nigeria officially implemented its National Photovoltaic Assistance Program Solar Home System (SHS), with the goal of benefiting some 25 million people.[19] The plan entails installing five million sets of solar home systems and small power grids nationwide for members of the population who lack power services or cannot connect to the grid. Beneficiaries can choose to pay weekly or monthly over 3 years based on the quantity of power generated. After 3 years, they become owner of their power generation systems. This result-oriented PSS design allows users to pay according to use. The SHS plan also involves manufacturing and assembly of photovoltaic components, helping local private solar companies obtain low-cost financing to provide services at low prices. The World Bank provided 20% of the funds for the SHS plan. Some Chinese photovoltaic panel manufacturers also participated by providing equipment support.

Philips' Lighting as a Service (Result-Oriented)
In 2018, Philips Lighting rebranded itself as Signify, marking a transition from selling products (mostly lightbulbs) to providing lighting services. Signify has implemented innovative services in collaboration with Cofely Energy Services at Schiphol Airport in the Netherlands.[20] This project is the world's first PSS design based on a circular economy, end-to-end model called Light as a Service (LaaS). According to this model, the airport pays only for the light it uses, while Signify retains ownership of the lighting fixtures and equipment, making it a typical outcome-oriented PSS design.

The project operates similarly to a product hosting model. Signify assumes all responsibilities for the design, installation, maintenance, and management of the lighting throughout its lifespan. The company ensures stable, lasting performance during the contract period, allowing customers to purchase lighting as a service

[19] Anugwom, E.E. (2022). Solar Home Systems in Rural Landscapes: Examining the Forces Shaping Solar Home Systems Adoption in Southeast Nigeria. In: Ojong, N. (eds) Off-Grid Solar Electrification in Africa. Energy, Climate and the Environment. Palgrave Macmillan, Cham. https://doi.org/10.1007/978-3-031-13825-6_9

[20] https://www.signify.com/global/lighting-services

rather than investing in new hardware upfront. Since Signify retains ownership and operational maintenance rights, maintaining maximum product value becomes crucial. To this end, the company has designed intelligent LED lighting fixtures and control systems that are modular and detachable for timely maintenance, upgrades, and repairs, ensuring long-term use. When their lifespan ends, Signify reuses, refurbish, or recycle the fixtures.

The intelligent control system precisely manages each fixture's operation based on Schiphol Airport's actual needs, existing lighting conditions, and personnel locations, thereby improving energy efficiency. Compared to traditional lighting, this approach offers improved optical performance, reduces energy consumption by 50%, extends the fixtures' lifespan by 75%, and significantly reduces environmental impact while saving on costs. Through highly customized services and continuous data collection optimization, LaaS reduces customer workload and provides a better lighting experience, establishing long-term trust relationships with customers.

9.5 Sustainability, Resistance, and Trends

If properly designed, PSS have great potential for promoting socioeconomic sustainability. From an environmental perspective, PSS incentivize manufacturers and service providers to reduce material and energy consumption throughout a product's lifecycle. Although each PSS project varies, manufacturers generally design more durable, energy-efficient products that are easier to repair, remanufacture, and reuse than they do otherwise. PSS design aligns with many of the design strategies discussed in previous chapters. From a social perspective, as opposed to selling material products, PSS focuses on environments of use. This approach can trigger active participation from local stakeholders, thereby increasing employment and promoting local economic development. Additionally, some PSS offer installment payment services, increasing the accessibility of these services to low-income groups by eliminating the need to purchase products upfront. Economically, PSS innovation can generate numerous advantages. Companies can access new market opportunities, enhance their competitiveness, increase the length and strength of their customer relationships, and create barriers for potential competitors.[21]

However, PSS also have limitations, such as possible rebound effects and increased maintenance, transportation, disposal, and recycling costs. The rise of the sharing economy, for example, has led to extensive high-frequency use and deliberate destruction of public goods, increasing the costs of product repairs and replacement and resulting in resource waste. Furthermore, PSS relies on infrastructure such

[21] Ceschin, F., Gaziulusoy, I. *Design for sustainability: a multi-level framework from products to social technical system*[m]. London: Routledge,2020:81.

"For companies, PSSs can, for example, offer the possibility to find new strategic market opportunities, increase competitiveness, establish longer and stronger relationships with customers and build barriers to entry for potential new competitors."

as networks, data, and logistics, which can increase the intensity of transportation and potentially generate additional carbon emissions and traffic problems.

Both businesses and users create the main obstacles to implementing PSS strategies. Companies accustomed to selling physical products need to change their organizational structures, business models, and philosophies. They must recognize that PSS are not about short-term profits but require long-term value realization, knowledge, skills, and talent reserves. Consumers, for their part, need to make a cultural shift from buying and owning products to using but not owning them.

Ultimately, PSS design is not merely a combination of products and services but an innovation to business and consumption models, reshaping the entire process of providing products and services. Its goal is to achieve benefits in terms of business value, human needs, and the environment. PSS represent a trend toward dematerialization, decoupling economic benefits from negative environmental impacts, and are one of the most promising design methods for sustainable development.

Chapter 10
Design for Social Innovation

Abstract Design for social innovation is a process of evolution for people, with people, and by people. Unlike the broader concept of social design, social innovation design has a clear value orientation. It aims to promote individual participation and cooperation through bottom-up approaches, fostering behavioral changes, reshaping social organizational mechanisms, cultivating social capital, and achieving social harmony and sustainable development.

Keywords Social innovation · Social sustainability · Participatory design · Social design

Sustainability issues revolve around relationships between humans and nature with the concepts, cultures, situations, and perspectives of various individuals, groups, and societies affecting those relationships. A lack of consensus among humans will lead to the "tragedy of the commons," threatening the existence of habitat we all share. This is the significance of social sustainability.

10.1 Social Sustainability

To address social sustainability, we must first understand what society is. The most fundamental concept in sociology, the term society, appeared in the fourteenth century, initially describing a group with common characteristics. Later, Durkheim defined society as an independent entity that has a wide, profound impact on individuals within a specific territory. In light of the rapid flow of material and personnel in the contemporary world, this definition based on nation-states has faced significant challenges.[1] The definition of society we use in this chapter is: a group of people who interact with each other over a long period of time, while sharing

[1] Giddens, Sutton. Essential Concepts in *Sociology* [M]. Translated by Wang Xiuxiao. Beijing: Peking University Press, 2019: 29.

X. Liu et al., *Design for Sustainability*, Sustainability and Interdisciplinarity,
https://doi.org/10.1007/978-981-96-8657-5_10

physical space or social territory often delineated by common political systems and cultural traditions.[2]

Compared to economic and environmental sustainability, social sustainability is a less discussed topic and lacks a universally accepted definition. In 1987, the UN published the "Report of the World Commission on Environment and Development: Our Common Future," in which the social dimension of sustainable development primarily denotes health and income equality. The 1992 "Rio Declaration on Environment and Development" included in that definition dignified living, justice within and between generations and internationally, and local participation in sustainable development. Current research on social sustainability mainly presents two perspectives: social development, measured by indicators like the UN's Human Development Index (HDI), which selects per capita GDP, average years of education, and life expectancy as the most representative parameters of social development, as well as high ecological efficiency and a small carbon footprint; and social cohesion, social harmony, and social capital, which remain controversial in the western social studies and difficult to measure.

In China, government officials and scholars rarely speak of social sustainability, but social development and the construction of people's construction are crucial national policy components. The Report of the 19th National Congress of the Communist Party of China addresses social development in terms of poverty alleviation, improving education in the central and western regions, income equity, social security, healthcare, affordable housing, social governance, and safety, emphasizing the goals of social justice and equal opportunity.[3]

Design with social sustainability as the goal emerged after design focused on environmental sustainability. This is partly because social problems often become apparent when economic development lags, and apart from the government, there is no clear responsible entity. Additionally, the complexity of social problems far exceeds that of specific products, spaces, or services.

Driven by a sense of responsibility, designers who are dissatisfied with practices aimed solely at generating commercial profits began engaging in social design. This field has evolved from expressing viewpoints to direct interventions, from providing specific solutions to complex systemic issues, and from interventions by a single design discipline to collaboration among multiple disciplines with practitioners maintaining open social participation throughout. Design practice aimed at innovating social organizational structures is becoming an increasingly important aspect of the design field.

However, innovating social organizational structures involves reconstructing complex social systems, and designers still need to deepen their understanding of these systems. Social innovation designers are still exploring and discussing its

---

[2] See entry for "Society" on Wikipedia (https://en.wikipedia.org/wiki/society)

[3] Xi Jinping. Winning the comprehensive construction of a moderately prosperous society and achieving a great victory in socialism with Chinese characteristics for a new era—Report at the 19th National Congress of the Communist Party of China [R/OL]. (2017-10-27) [2021-09-16]. http://www.gov.cn/zhuanti/2017-10-27/content_5234876.htm

connotations, recognizing that its methods and methodologies are far from mature. Nevertheless, various notable cases provide insights into the core issues under debate and highlight the goals designers can achieve with their current professional abilities.

10.2 Social Design: Design for the People

Early design interventions into social issues often tackle recognizable problems by proposing specific solutions, such as providing affordable products and services for the poor's basic survival needs. For complex issues without straightforward solutions, designers express their attitudes through design or art actions, acquiring social influence and prompting solutions. Both types of design practices share inherent similarities: one is designer led and the other uses relatively traditional design methods.

10.2.1 Design for the Bottom of the Pyramid

Scholars often trace the term social design back to Victor Papanek, although he never directly proposed it. In 1991, Nigel Whitely introduced the term in *Design for Society*. Although Whitely admits that his simple title can easily lead to misunderstandings and disputes, he believes it is necessary to ask whether consumerism-oriented design should dominate design, as well as whether consumerism design is beneficial or harmful to society as a whole. Whitley writes that it is necessary to continue discussing society as a concept: is society equal to the sum of all individuals?[4] He likewise proposes that green design, design for vulnerable groups, design in line with consumer ethics, feminist design, and so on are all aspects of design for society.

Both Papanek and Whitely advocate for design for vulnerable groups, also called design for the bottom of the pyramid (BoP). This approach reflects a sense of social responsibility shaped by liberalism, which advocates for individualistic values with personal freedom being the highest value, and neoliberalism, a close relative to social Darwinism that generates acceptance for social stratification. When intellectuals and designers like Papanek view stratification from a humanitarian perspective, they embrace BoP design.

BoP design focuses on the basic survival needs of ordinary people, following Maslow's hierarchy of needs, such as food, water, energy, sanitation, and transportation. This makes it distinct from mainstream design that serves the purchasing power of the middle and upper classes. In recent years, the German Red Dot Award,

[4]Whitely N. *Design For Society* [M]. London: Reaktion Books, 1993.

the IF Design Award, and Japan's G-Mark have all included social design or design with a sense of social responsibility in their evaluation criteria.

LifeStraw Water Purifier
Drinking water infrastructure in African countries is limited. According to the World Health Organization, 855 million people lacked clean water in 2015, most of whom were in Africa. LifeStraw, a portable water purifier developed by Vestergaard Frandsen, filters out 99.9999% of waterborne bacteria, providing clean water. Priced at $25, it can be used by one person for a year, filtering 750 liters of water without the need for replacement parts. Sold to local NGOs, the product reaches end-users for $10. Vestergaard Frandsen has provided 60,000 water purifiers to Lutheran World Relief, as well as to 482 clinics in the Democratic Republic of Congo and 5000 residents in Niger. According to statistics, LifeStraw has benefited two million children through a humanitarian feedback program.

10.2.2 Design for Social Impact

Design for social impact has multiple definitions in English and is often used interchangeably with terms such as social design and social innovation design. In this chapter, we define social impact design as targeting specific social issues and using design in installations or actions to express attitudes towards these issues. The goal is to garner widespread social attention and ultimately promote their resolution. Having an impact is the main goal of such design, making social communication a crucial component. By means of media, visually impactful, participatory installations attract public attention and participation, express the designer's views, and attain support and consensus from the general public.

Social impact design bears a close relationship to other emerging schools of design like critical design, speculative design, and design activism. These schools likewise focus on social issues such as environmental pollution, social segregation, immigration, alienation due to technology, and gender. Designers working in these areas seek to express their views on issues rather than propose solutions, with projects that are often installations rather than mass-producible products or services. The effectiveness of these designs manifests in the degree of buzz and attention they generate. Compared to design for the bottom of the pyramid, which often generates specific solutions, social impact design emphasizes expression of concepts.

The Teeter-Totter Wall
The issue of immigration on the US-Mexico border has had a long-lasting, complex impact on American society. Ronald Rael, a professor of architecture at the University of California, Berkeley, and Virginia San Fratello, an associate professor of design at San Jose State University, created an installation called Teeter-Totter Wall. After President Trump pushed for erecting a wall along the entire length of the US–Mexico border, the two designers installed three pink teeter-totters on the wall in the Anapra area of Juarez City, Mexico, hoping to give rise to discussions about

the border issue in a very candid and humorous way. Children and adults from both sides of the border participated. Although the installations only existed for 20 min before the police dismantled them, videos of people playing on the teeter-totters went viral online.

The Pink Teeter-Totter won the 2020 Beazley Design of the Year award from the Design Museum in London. Tim Marlow, CEO and director of the museum, remarked: "The wobbly wall encourages new ways for humans to connect. It is a creative work and will always play the role of a reminder, reminding us how humans can transcend those forces that divide us."[5]

10.2.3 Limitations of Social Design

Whether addressing specific problems or raising awareness, social design revolves around specific issues, proposing solutions or expressing the designer's attitude to an issue through the designed object. In the context of prevailing consumerism, such independent thinking and socially responsible design practice are precious. However, increasing numbers of researchers and practitioners recognize that complex social problems, such as wealth disparity, the difficulties of aging, and immigration, exceed the capabilities of a single design discipline. Design interventions, whether they provide a spark to certain topics, initiate dialogue, or result in products or services, have limited impact in terms of solving complex social problems. Such problems are deeply rooted in society, and without local resources and largescale popular involvement, design solutions often fizzle out after the prototype stage. Converting design for social impact into calls for action is a struggle, and after initial enthusiasm among a limited set of people, the root problem remains. This may even lead to public doubt about the design discipline's capabilities.

The ability of design practice to contend with real-world problems is a core challenge to social design. Designers need to possess a system view and more humanities knowledge than they have. They must understand that complex problems require multidisciplinary collaboration and co-construction with the participation of different stakeholders. During such collaborations, making the best use of professional advantages and changing elitist attitudes and methods is crucial to the future development of design as a discipline.

[5]Pink Seesaws At US-Mexico Wall Win Design Award[EB/OL]. (2020-01-19) [2021-09-11]. https://www.bbc.com/news/world-us-canada-55718478

10.3 Participatory Design: Design with the People

When designers address non-market-oriented problems, whether that entails proposing solutions or expressing clear attitudes, they absorb new concepts and methods into their discipline. Interaction and collaboration with users have become key to design innovation. Participatory design is essential to social innovation design because without social participation, the goals of social innovation design cannot be achieved.

10.3.1 The Rise of Design Democratization

As an activity that involves creation, the discipline of design has been accumulating teachable methods since the handicraft era. These include basic material and color studies, as well as structural and form studies that draw on engineering. The Bauhaus movement systematized these methods, creating the basis for contemporary design.

After World War II, with the rapid development of modern industry in the West, design departments and independent designers continuously introduced innovations to design methods. In this period, American industrial designer Henry Dreyfuss, considered the founder of ergonomics, focused on product comfort and extensively studied ordinary people's basic physical dimensions, abstracting from these rules and standards for product design. Large consumer goods companies established research departments, recruiting users to test devices for which their researchers extensively borrowed experimental methods from psychology and sociology. Using face-to-face interviews, observations, user scoring, and even physiological index tests such as electrocardiograms, eye movements, and brain waves, designers evaluated user interest and excitement, developing products with unique selling points.

These design research methods have achieved great success in the manufacturing industry. However, when designers address social issues, it becomes challenging to apply these methods. First, design aimed at solving social problems often entails services, spaces, and systems rather than material products. Thus, design methods based on individuals' reactions to products have limited applicability. Second, the end-users of such designs are diverse, and the stakeholders are complex. For instance, public space users may include the elderly, children, tourists, families, and groups of young people. Their interests and behavior patterns may vary widely and even come into conflict with each other. Laboratory-style research methods often strip away specific situations, making it difficult to capture real needs and design opportunities. Third, governments implement social designs such as public objects, facilities, and services after long-term planning and budgeting, and therefore enjoy natural monopolies. Once implemented, they often leave users with no other choice. Moreover, modifying designs after problems arise is challenging, making the need for high-quality preliminary design research all the more crucial.

Participatory innovation, where users directly participate in the design process and propose design concepts, emerged in the field of northern Europe's information systems in the 1970s and was later widely applied in architecture and planning.[6] In the twenty-first century, participatory design that engages with social issues has become a major focus of design innovation, especially with the rise of information and interaction technologies.

10.3.2 Community Design

Communities are central to participatory design. Like the term society, community is a core concept in sociology, containing two elements: sharing a physical space, although the existence of virtual communities challenges this aspect of the definition, and sharing values and interactions. A community is smaller and easier to comprehend than a society.

The methods and processes of community design have a certain measure of replicability. Community design focuses on physical traits, such as facilities and spaces, while community-based design focuses on spiritual ones by providing comprehensive solutions to groups that share similar values and have high rates of interaction. In either case, extensive participation among residents is indispensable to achieving long-term support.

Community design is a significant practice in Europe and Japan. In recent years, in China, with urbanization moving from incremental development to stock renewal, community design practices have aimed to improve residents' quality of life in old urban areas have been thriving. Community public spaces and community gardens have become carriers of participatory design projects, with localized transformations leading to the formation of a design methodology with Chinese qualities.

Cenni Collective Housing in Milan

In 2006, Politecnico di Milano launched a master's program in Collaborative Housing Design to cultivate students' abilities to assist residents to organize, design, and build residential communities independently. Following graduation, Ph.D. student Giordana Ferri, who specialized in collective housing, established the Social Housing Foundation (*Fondazione Housing Sociale, FHS*). Between 2012 and 2015, Ferri completed the largest social housing project in Europe, the Milan Cenni Collective Housing community. Ferri's process of design intervention exemplifies the fundamental concepts and methods of social innovation design.[7]

Social housing significantly differs from conventional housing. Traditional commercial and public rental housing prices include the property itself and associated

---

[6] Zhong Fang, Liu Xin. Design for the People, by the People, and of the People: The Path, Challenges, and Opportunities of Social Innovation Design[J]. *Decoration*, 2018(5): 40–45.

[7] Altreconomia Fondazione Housing Sociale. *Il Gestione Sociale: Amministrare Gli Immobili E Gestire La Communita Nei Progetti Di Housing Sociale*[M]. Milano: Altreconomia, 2011.

services, such as security, public health, waste disposal, and maintenance. The quality of these services correlates directly to their cost. In contrast, social housing project developers implement property services based on residents' needs and manage them through resident involvement. This active participation and frequent interaction foster close interpersonal relationships within the community. High-quality living spaces, personalized community services, and strong community relationships form the unique essence of social housing, especially the latter, which cannot be generated through market transactions. Social housing encompasses three dimensions—space, service, and community relationships—each undergoing a thorough process of demand definition and product development.

The first step in a social housing project is financial planning. The Milan Cenni Collective Housing project first established an investment committee to evaluate the project's cost–benefit and issue low-interest loans to buyers. User recruitment, product development, and project construction followed.

The first stage involved establishing mechanisms for recruiting and selecting residents. The Social Housing Foundation launched an online platform to attract individuals in need of social housing and willing to participate in the complete design and construction process. The aims of the project organizers included creating a diverse community, prioritizing families with urgent housing needs and low-to-middle incomes, thus pricing units relatively low. The condition for participation was agreement with the community concept and a willingness to engage in community life. The resulting composition of residents was diverse, including multigenerational families, nuclear families, young single people, families undertaking social obligations (such as adopting orphans or providing temporary housing for people in crisis), elderly people, and foreigners. Residents formed a management team, the functions of which greatly exceeded those of conventional property management. These included spontaneous facilitation of community activities, providing community services, enhancing interpersonal interactions, and promoting transformation of residential communities into cohesive entities with shared values and stable social networks.

The second state entailed development of the hardware of the living space. In conventional housing projects, developers pre-design living spaces and outdoor landscapes, leaving consumers to choose based on their needs. Social housing, however, employs a bottom-up custom-design process. After categorizing various demands, designers proposed three housing types—small, medium, and large—to accommodate 10 different family structures and functional requirements. This process involved extensive participatory design, allowing residents to ultimately determine the product form, including the choice of building materials.

The third stage is the development of the software of community services, the most distinctive and socially valuable quality of social housing. Through a series of participatory design workshops, designers used tools like demand cards and scenario boards to stimulate residents to propose needs and integrate them into the architectural design. Consequently, the Cenni residential complex introduced over 10 types of community services, including family nurseries (inspired by the EMUDE

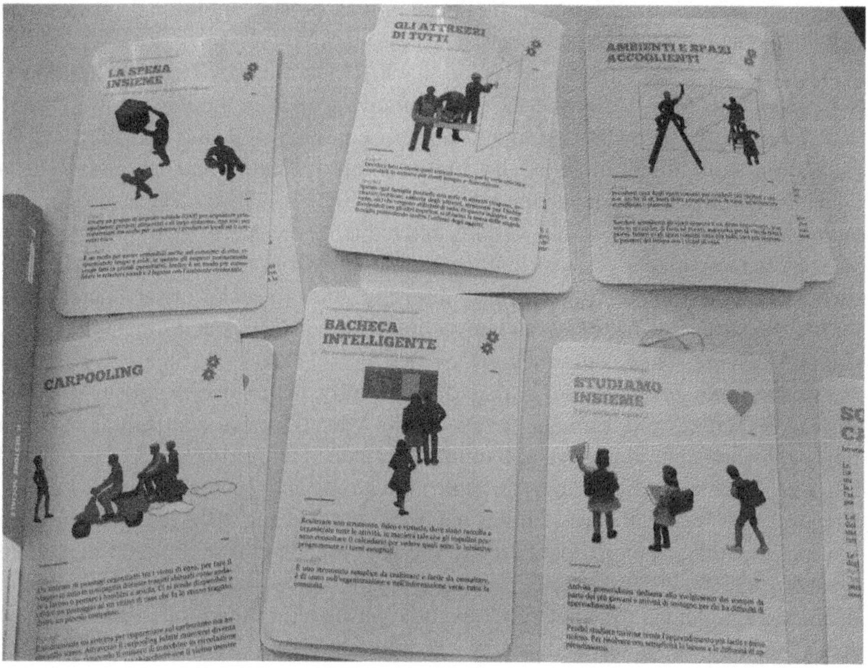

Fig. 10.1 Collaborative service cards for participatory workshops (Bardelli, G., Ferri, G., Pavesi, A. S. *Starting Up Communities: Un Design-Kit Per L'abitare Collaborativ* [M]. Milano-Torino: Fondazione Housing Sociale, 2016.) (Photo by author)

Project), community vegetable gardens, children's painting rooms, shared kitchens, cooking classrooms, community libraries, and time banks.

After several years of design, construction, and operation, the Cenni residential project established a comprehensive methodology. Its innovative workflow, tools, and methods have been practically validated (Fig. 10.1) and continue to be iterated in new projects.

10.4 Design for Social Innovation: Design by the People

10.4.1 Background on Social Innovation

Michael Young, the founder of the Open University in the UK, first proposed the concept of social innovation in the 1960s. However, social innovation only gained public prominence in the early twenty-first century. Western sociologists refer to the government, business, and society as the first, second, and third sectors, respectively. Social innovation is linked to civil society as part of the third sector, representing innovation initiated by social forces outside traditional government and business.

The rise of social innovation traces back to the Reagan-Thatcher neoliberal reforms of the 1980s. After the 1970s, economic growth in the West slowed, there was an energy crisis, and developed Western countries fell into the quagmire of stagflation with high inflation and high unemployment. Thatcher and Reagan launched neoliberal reforms in the UK and the US, mainly by cutting taxes, deregulating, drastically cutting social welfare, and reducing the size of government. These reforms led to over 30 years of high economic growth in the US. However, by the early twenty-first century, particularly after the 2008 financial crisis, social problems in Western countries emerged, revealing severe insufficiencies in public goods and intensifying social contradictions.

At the same time, countries that did not fully adopt neoliberalism likewise entered a similar predicament. After World War II, most continental Western European countries adopted a system that guaranteed a high standard of living and relatively low wealth disparity through high taxes and generous social welfare. However, the financial crisis caused many countries, especially those in southern Europe, to endure national debt crises. The tax revenue that supported the welfare system diminished considerably causing a drastic reduction in public expenditure. At the same time, aging populations and new waves of immigration introduced new social challenges, likewise creating a situation where social contradictions intensified.

People in Western countries began to realize that their existing methods were inadequate for coping with profound social problems and came to regard social innovation as a third way beyond big government and an omnipotent market. In this context, organizations such as the British Council and the Young Foundation promoted a large number of social innovation projects and the Obama administration in the United States established a Social Innovation Office in 2009.

10.4.2 Social Innovation

The most commonly cited definition of social innovation is: social innovation refers to new ideas that work in meeting social goals.[8] This definition makes the best sense in the context of new solutions outside traditional public welfare and marketization models. For example, traditional elderly care models—family self-care, public nursing homes, and commercial services—all have shortcomings. Low- and middle-income families may lack the ability to pay and be unable to afford full-time care and relatively high-cost commercial services; public nursing homes require significant government financial input and long-term planning. Otherwise, they

[8] Mulgan, G., Tucker, S., Ali, R., Sanders, B [M]. *Social Innovation: What It Is, Why It Matters And How It Can Be Accelerated.* London: The Basing Stoke Press, 2007.

cannot accommodate more elderly people and meet social needs. Seeing these challenges, some local government officials and residents in Europe, where aging is a significant issue, have proposed new community service models that incorporate government agencies, communities, and residents, resulting in innovative social innovation models.

One example is the Prendi in casa uno studente (Take a Student Home) project in Italy, which has one of the largest aging populations per capita in the world. In Milan, Italy's largest city, not only has a large aging population but also an increasing proportion of single-person households. At the same time, Milan's housing prices are high. Universities cannot provide affordable dormitory accommodation to students, which means many students have to live in surrounding cities and commute to school every day. The civil society organization Megliomilano (Better Milan) initiated a project to address high housing costs for students and loneliness among the elderly by pairing elderly people with spare rooms with university students in need of affordable housing. Students pay half the market rent, providing basic daily care and companionship in return. The project involved psychologists for assessments and pairing, ensuring mutual trust and cooperation. From 2004 to 2014, the project successfully facilitated over 650 agreements, with only eight failures.[9]

In this case, the government, society, and citizens all invested resources and shared costs. For example, the government subsidized pensions for the elderly; social organizations were responsible for assessments and pairing, and provided daily support and guidance to the elderly and students; the elderly provided empty rooms; young students provided time and energy, ultimately achieving a win-win solution for the government, the elderly, and the students. This solution required mutual trust and cooperation between the elderly and students, bolstered by Italy's strong family tradition and a model of multi-generational cohabitation. In other regions where individualism is more common, it would be difficult to replicate this model.

New solutions beyond the government and market depend on the organizational capacity of society itself, that is, social capital. Social capital, a highly debated concept in sociology, refers to the value individuals recognize in social networks, reciprocated through these networks.[10] Shared values, social networks, interpersonal trust, reciprocity, and cooperation are core elements of social capital. In the past 30 years, social capital has also been a hot research field in sociology, with important research published by Bourdieu, Putnam, Fukuyama, and others. Many scholars believe that social capital directly affects the efficiency of the government, ultimately affecting social and economic development levels.[11] Social innovation

---

[9] Rewritten from: *Manzini. Design,* When Everybody *designs: An introduction to social innovation design* [M]. Zhong Fang, Ma Jin, trans. Beijing: Electronics Industry Press, 2016.

[10] Bourdieu P. *The Forms Of Capital*[M]; Richardson J G. Handbook Of Theory And Research For The Sociology Of Education. New York: Greenwood Press,1986: 241–258.

[11] Putnam. *Making Democracy Work: Civic Traditions in Modern Italy*[M]. Wang Lie, Lai Hairong, trans. Beijing: Renmin University of China Press, 2015. Robert D. Putnam, Robert Leonardi),

often begins with human connections that build trust and cooperation over the course of long-term interactions, shaping common values. Replacing traditional business capital and government investment with social capital may consequently give rise to numerous innovative social organizations and economic models.

10.4.3 Methodology of Design for Social Innovation

In 2004, Professor Manzini's team from Politecnico di Milano launched the European Commission's research project, Emerging Users Demands for Sustainable Solutions (EMUDE). The project researched cases of social innovation in the EU. In 2007, it published "Creative Communities," featuring 59 cases spanning six themes: living, eating, commuting, working, learning, and socializing.[12] Nearly 20 years later, some practices remain innovative, while others have become widely accepted, including carpooling, shared bicycles, couchsurfing, organic farming, and collective buying.

Collecting and organizing real-world cases provides valuable inspiration for social innovation design. By tracking, understanding, and analyzing these cases, researchers can identify similar patterns and implementation strategies, thus facilitating deliberate design interventions.

In 2009, Professor Manzini initiated the Design for Social Innovation and Sustainability Network (DESIS), which currently includes 52 design colleges from five continents, including six in China. The design institutions in this network follow common principles, collect information about design innovations in their regions, and exchange and share data, promoting localized social innovation design practices while also providing knowledge and experience potentially applicable to other countries and regions. Summarizing diverse patterns from different social backgrounds and intervention fields yields basic common methods:

1. *Building social connections between people around specific themes and promoting interpersonal interactions*: The foundation of social innovation is social capital, which does not arise naturally, particularly in modern societies characterized by individualism and atomization.[13] Hence, establishing positive interpersonal interactions is often the first goal of designing interventions for social issues. This emerges from creating community activity spaces, community gardens, and community libraries, which attract members of the population. The next step is to design physical and virtual touchpoints (images, facilities, places, and online platforms) to stimulate interpersonal interactions.

Raffaella Y. Nanetti, Making Democracy Work: Civic Traditions in Modern Italy

[12] Meroni, A. *Creative Communities: People Inventing Sustainable Ways Of Living* [M]. Milan: Edizioni Poli. Design, 2007.

[13] Putnam, Robert D. Bowling Alone: *The Collapse and Revival of American Communities* [M]. Translated by Liu Bo, et al. Beijing: China University of Political Science and Law Press, 2018.

2. *Creating and depicting visions to stimulate participation by ordinary people:* To stimulate participation, it is crucial to create scenarios that encourage ordinary people to engage in innovation. Scenario building and visual, auditory, and other communication exercises can effectively stimulate creativity and action. Scenarios often draw on real-life experiences, which, using conventional design methods like cutting, pasting, and role-playing along with certain spaces, environment, and characters, can elicit the transformation of abstract concepts into local, concrete situations.
3. *Making users part of social solutions:* In the absence of commercial capital and government financing, implementation and long-term operation of innovative concepts must rely on the will and abilities of ordinary people. These form the basis for social innovation project sustainability. This, in turn, relies on designers' deep insights into project participants and stakeholders, as well as reconstruction of the social organizational structure. Such reconstruction requires collaboration with community workers, informal groups, and understanding of various policies.

10.4.4 Challenges of Social Innovation Design

In social innovation design, the role of designers fundamentally differs from their traditional roles.

First, designers are no longer the subjects of creativity, but instead refine, integrate, and replicate the claims and creativity of ordinary people. This shift requires significant transformation and updates in both concepts and methods among professionally trained designers.

Second, the process of social innovation design is more complex. Social innovation design involves an open, cooperative process from problem identification to project completion and operation. Designers must be involved throughout the process, which is more complex than traditional design methods that involve design acceptance and scheme freezing. Understanding and negotiating the demands of numerous stakeholders, helping different groups and individuals reach consensus, requires skills more akin to social work, which are not typically included in current design education.

Finally, social innovation design places high demands on the designers' awareness of problems, social cognition, and social organization capabilities. Unlike traditional design problems, which arise as commissions, social innovation challenges depend on designers' insights. Cultivation of this ability exceeds the scope of traditional design education, requiring active acquisition of interdisciplinary knowledge, especially with regard to acquiring a deep understanding of social issues. Proposing solutions also relies on discovering social capital and excellent social cooperation skills.

10.5 Conclusion and Reflection

Compared to environmentally sustainable design, social innovation design is more directly influenced by social context. The sociology underlying social innovation originated in the West. All the same, indigenous Chinese sociological theory has been growing, constructing frameworks that align with Chinese reality, history, and culture. This includes new understandings of the relationships between society, the state, and the market.

Research and practice in social innovation design require a profound understanding of the similarities and differences in social organizations and cultures across different contexts. Exploring design methods and theories with local characteristics is essential to this process. Social innovation design places people at the core of design issues, creating a process that is for people, with people, and by people. It aims to promote awareness and cooperation from the bottom up, leading to behavioral changes and reshaping social organizational mechanisms.

If environmentally sustainable design focuses on reducing environmental impact with the product at the center, socially sustainable design aims to solve real problems through people's initiative, meet real needs, and achieve social harmony and prosperity. However, the practical implementation of social innovation design is highly dependent on the foundation of social capital in specific communities, which is a fundamental social problem, not just a design issue. Therefore, ongoing collaborations with disciplines such as sociology and public management is essential to explore methodologies suited to local social and cultural backgrounds. As a practical discipline, design must continuously attempt and explore to foster social innovation effectively.

Chapter 11
The Distributed Economy

Abstract The distributed economy is an emerging, highly sustainable economic model that tackles a range of complex contemporary and future challenges. It does so by establishing a production and consumption system based on localized, decentralized, open small-scale production units. By deeply understanding both the potential and limitations of this model, designers can play an active role in promoting the sustainable transformation of the world's socioeconomic landscape.

Keywords Distributed economy · Decentralization · Distributed manufacturing

This chapter delves into the historical origins and fundamental concepts of distributed systems. It explores the different types and characteristics of the distributed economy and demonstrates its immense potential for contributing to sustainable development and the construction of an ecological civilization, supported by a series of design cases.

11.1 History of Distributed Systems

The concept of distributed systems traces its origins to 1964 when the American RAND Corporation first proposed it. Initially applied to a distributed communication network developed for the US military, the primary goal was to enable military forces to respond rapidly, efficiently, and flexibly to large-scale attacks, both online and offline. The RAND Corporation sought to replace the traditional centralized system with a decentralized system that allows all nodes in the network to share information equally, which evolved into the distributed system (Fig. 11.1).

As internet technology advanced, the concept of distributed computing emerged and was widely adopted in natural science research. This approach involves dividing large engineering datasets into smaller blocks, which multiple computers then process independently. The results are subsequently unified to produce final conclusions. Modern distributed computing projects often leverage the idle computing

X. Liu et al., *Design for Sustainability*, Sustainability and Interdisciplinarity, https://doi.org/10.1007/978-981-96-8657-5_11

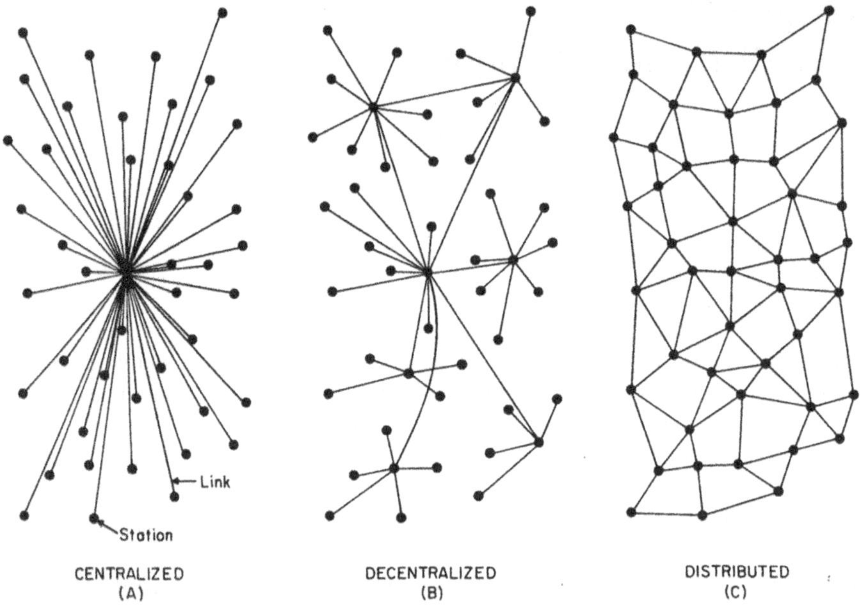

Fig. 11.1 Centralized, decentralized, and distributed networks (Baran, P. On distributed communications: i. Introduction to distributed communications network[R]. Santa Monica, CA: Rand Corporation, 1964: 2)

Fig. 11.2 Folding@home Project uses distributed computing (Image source: Wikipedia)

power of millions of volunteer computers worldwide, transmitting data via the internet. Examples include the Folding@home Project, which analyzes protein structures and drug components, and the Climateprediction.net Project, which simulates historical and future global climate patterns to address catastrophic weather events (Fig. 11.2).

In addition to scientific research, distributed systems have applications in other fields, including gaming. For instance, in 2018, the scientist Devin P. Sullivan, working with others, used publicly available cell atlas data from the Human Protein

Atlas (HPA) in the massive, multiplayer game EVE Online. They integrated an image classification task into a game module called Project Discovery, where players were asked to identify and classify protein staining patterns in human cells. Over the course of a year, 322,006 players contributed nearly 33 million subcellular classifications, including many new categories not previously recorded. The results were made freely available to scientists worldwide.

In the above cases, most of the contributors were not professional researchers; rather, they participated in these large-scale collaborative efforts out of personal interest or to complete tasks in a game. This unique distributed model is characterized by openness, spontaneity, efficiency, collective wisdom, and equal interaction. These traits have since appeared in various other social and economic fields.

For example, the renowned game development company Valve, founded in 1996, has released a series of popular products including Dota 2, Half-Life, and the digital game distribution platform Steam, with over 20 million users worldwide. What sets Valve apart is its entirely flat, distributed management model. Unlike traditional top-down management structures, Valve allows employees to spontaneously form project teams, publish tasks, and recruit team members through internal networks or cross-group meetings. Employees then choose to join teams based on the popularity of the project, as well as their own needs, interests, and talents, creating a dynamic balance between employees and teams. Despite the absence of fixed job scopes and restrictions, Valve's employees clearly understand their responsibilities, which they define through communication with colleagues and team members. Moreover, all of Valve's teams form on a voluntary basis, with each member taking responsibility for their participation and self-management.

Allan Johansson and his colleagues from Lund University argue that the focus in modern industry on efficiency and output has resulted in the creation of centralized, large-scale production units which a few powerful business institutions control. This concentration of production has led to the extensive consumption of non-renewable resources and significant waste generation. Simultaneously, a reduction in transportation costs over time has facilitated the long-distance transportation of raw materials and products, concealing the true social and environmental costs of production. This dynamic has, to some extent, distorted and eroded local cultures, limiting the development of regional economic diversity. In response, in 2005, Johansson and his team proposed a small-scale, dynamic, self-organizing business model for the European context, which they termed distributed economies (DE). They envisioned this concept guiding the current industrial system toward a sustainable future.[1]

---

[1] Johansson, A., Kisch, P., Mirata, M. Distributed Economies-A New Engine For Innovation[J]. *Journal Of Cleaner Production*, 2005, 13(10–11): 971–979.

11.2 Concepts, Characteristics, and Types of Distributed Economy

11.2.1 Concept

The distributed economy is not a utopian ideal emerging from nothing. Historically, economic models in human society evolved from decentralized systems to centralized ones, and then from centralized systems to distributed structures (Fig. 11.3). Early decentralized production models, such as workshops, catered to individual needs but were inefficient and fragile. The centralized production model, which large factories exemplify, marks a significant achievement in human social development and collaboration, characterized by efficient mass production. However, this model often erases individuality and incurs high system maintenance costs. Over time, its ability to correct errors and adapt to changing circumstances has manifested significant flaws. Decentralization emerged in response to the limitations of centralized systems, as a way to explore new models that can potentially balance individual needs and collective production.

Distributed systems seek to meet individual needs in localized contexts, revitalizing local economies while also considering environmental impact, production efficiency, social resilience, and individuality. This concept calls for a transformation in the industrial system towards DE, moving away from the socioeconomic and environmental challenges inherent in large-scale, centralized production favored by neoclassical economic drivers.

According to Johansson, the distributed economy could potentially evolve into a new socioeconomic structure that better supports the growth of localized, small-scale enterprises. The International Institute for Industrial Environmental Economics (IIIEE)[2] and projects such as the Sustainable Design International Learning Network

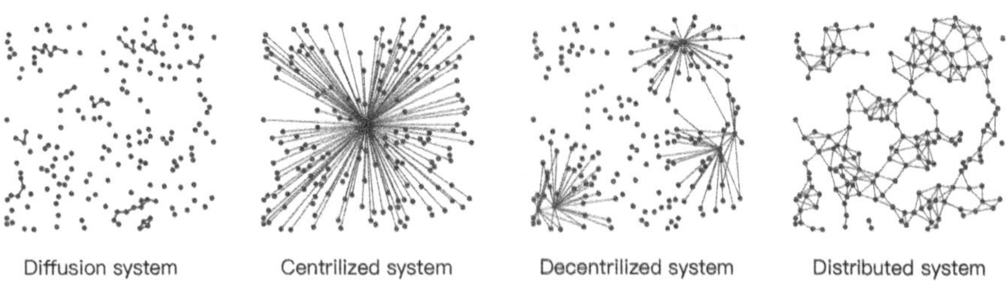

Fig. 11.3 Four system models (Massimo, M. A Framework For Understanding The Possible Intersections Of Design With Open, P2P, Diffuse, Distributed And Decentralized Systems[J]. *Journal Of Design Culture*, 2016(1–2): 44–71) (Image source: Drawn from the source in footnote 1)

[2]The International Institute for Industrial Environmental Economics (IIIEE) is a research institution affiliated with Lund University in Sweden. It focuses on research in the fields of sustainable consumption and production, sustainable product service systems, and sustainable energy and building development.

Table 11.1 Definitions of the distributed economy

Authors	Year	Definitions
Allan Johansson, Peter Kisch, Murat Mirata	2005	DE entails distribution of a selective share of production to regions where small-scale, flexible units, which are synergistically connected with each other and prioritize quality in their production, organize a diverse range of activities.
IIIEE	2009	Distributed economies [are] an alternative structure for society and economy with rather small-scale businesses in a local economy context, leading potentially to a more sustainable social and economic structure.
LeNSin	2016	Distributed economies (DE) could be defined as small-scale production units, at or near the point of use, where the users are the producers—whether individuals, small businesses, and/or local communities. If the small-scale production units are connected with each other to share various forms of resources (physical and human knowledge-based ones; for example, to share the energy surplus), they become a locally distributed economy network, which may in turn be connected with nearby similar networks. If properly designed, they are promising to promote locally based social equity and cohesion.

(LeNSin)[3] have further developed this idea and provided a comprehensive definition of the distributed economy (Table 11.1).

11.2.2 Characteristics

The key to understanding the distributed economy, as an economic model built on a distributed system of social technology,[4] lies in understanding the system's nodes and the relationships between them. In a distributed economy, a node is a small-scale production unit, which can be an individual, a local enterprise, an institution, or an organization. In the ideal model, all nodes can fairly obtain natural resources and social resources, information, and knowledge and share them with other nodes in the network. In short, the distributed economy aims to stimulate and promote the vitality and development of local, small-scale, decentralized, flexible production units, thereby driving the development of the overall social economy (Fig. 11.4).

Comparing the relationships between nodes in the models of different system reveals the smallness of the distributed system's scale and that the structural relationship between each production unit is more equal, compared to the other two

[3] The European Commission (ERASMUS+ Higher Education Capacity Building Project) funded and administered LENSIN (The International Learning Network of Networks on Sustainability) from 2016 to 2020; LENSIN, based on the original LENS project (2007—2010), incorporated 36 universities worldwide as primary partners to collaborate on the theme of "Sustainable Product Service System Design and Distributed Economy." See http://www.lens-international.org.

[4] For an explanation of social technical systems, see Chapter 2.

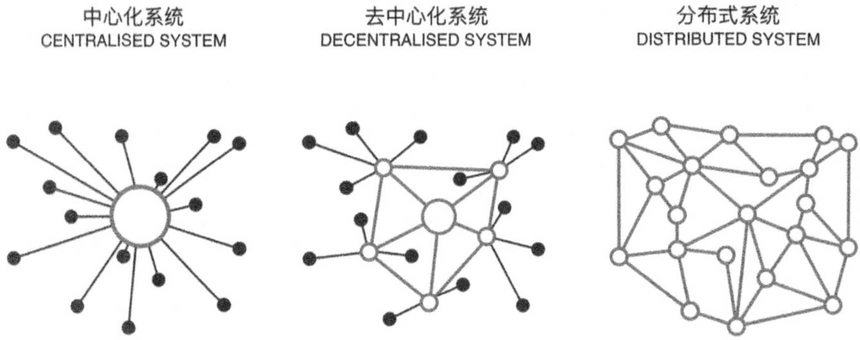

Fig. 11.4. Relationship between nodes under different system models (International Institute for Industrial Environmental Economics (IIIEE). (2009). The Future is Distributed: A Vision of Sustainable Economies. Lund: IIIEE: pp 2)

Table 11.2 Comparing centralized systems to distributed systems

	Centralized System	Distributed System
Size	Large	Small
Structure	Hierarchical	Heterarchical
Responsiveness	Static	Flexible
Resilience	Vulnerable	Resilient
Proximity to Users	Distant	Close
Openness	Closed	open (hybrid)
Diversity	Standardized	Customized

models. In response to external changes, the system is not only more resilient than the others, but also more flexible and proactive. Compared to a centralized system, the production units in a distributed system are closer to users and more open, which can motivate users to actively participate, develop customized solutions, and effectively meet individual needs (Table 11.2).

The development of the distributed economy is intrinsically linked to the combined forces of social and technological innovation. Viewed from a socio-technical systems perspective, the distributed economy represents an integration of social innovation and technological change, which permeates various levels of the socio-technical system by transforming both infrastructure and production-consumption systems.[5] Just as the internet and digital technology support mobile payments, thereby fostering the sharing economy, the distributed economy will also thrive under the impetus of a new round of technological innovation.

[5] Manzini, E., Coad, R. *Design, When Everybody Designs: An introduction to Design for Social Innovation*. MIT Press, 2015: 16–17.

11.2.3 Types

The Sustainable Design International Learning Network (LeNSin) categorizes the distributed economy into seven types, each based on the different resources shared by production units:

1. Distributed Manufacturing (DM): Includes localized workshops utilizing technologies like 3D printing.
2. Distributed Food Production (DFP): Encompasses activities such as urban farming.
3. Distributed Water Systems (DWS): Involves systems like rooftop rainwater collection.
4. Distributed Renewable Energy (DRE): Comprises small-scale, localized solar and wind power systems.
5. Distributed Software (DS): Refers to open-source software like Linux.
6. Distributed Information/Knowledge (DI/DK): Examples include platforms like Wikipedia.
7. Distributed Design (DD): Includes networks such as Rococo and Zhubajie.

These classifications reflect the current state of development within the distributed economy. As technology advances and ideas proliferate, these types will expand and become more subdivided. In fact, the concept of distribution has already begun to integrate itself into various sectors of our socioeconomic landscape—including into finance, healthcare, and urban planning—forming a significant undercurrent shaping both our present and future (Table 11.3).

11.3 Sustainability of the Distributed Economy

Compared to the centralized economic model, a distributed economy offers significantly greater potential for sustainability. Harnessing this potential to advance sustainable development across society is a critical responsibility for designers at present. The following sections examine the sustainability potential of the distributed economy from environmental, social, and economic perspectives, providing guidance and support for design interventions.

Environmental Dimension: The distributed economy aligns production units, local resources, and end users as closely as possible, enabling the optimal use of localized resources, including raw materials and renewable energy sources like solar, wind, geothermal, and hydro power. This alignment significantly reduces energy consumption and the carbon emissions associated with long-distance transportation during production and consumption. Additionally, it can mitigate traffic congestion and noise. The distributed economy also enables the establishment of more environmentally friendly, energy-efficient, and intelligent small-scale production units.

Table 11.3 Primary research on the distributed economy and its types

Researchers	Year	Field	Publication
Allan Johansson, Peter Kisch, Murat Mirata	2005	Distributed Economics	*Distributed Economies—A New Engine for Innovation* Introduces the concept of distributed economy and its characteristics.
Mirata Murat, Helen Nilsson, Jaakko Kuisma	2005	Distributed Economics	*Production Systems Aligned with Distributed Economies: Examples from Energy and Biomass Sectors* Thoroughly discusses how locally based, small-scale, flexible production units can benefit local environmental and economic development.
IIIEE (Lund University)	2009	Distributed Economics	*The Future Is Distributed: A Vision of Sustainable Economies* Detailed case studies on distributed economics.
Che Biggs, Chris Ryan, John Wiseman, Kirsten Larsen	2009	Distributed Water Systems	*Distributed Water Systems: A Networked and Localized Approach for Sustainable Water Service* Discusses the trend towards distributed urban water systems and proposes the core structure of a distributed water system using case studies.
Bianchini Massimo, Stefano Maffei	2013	Distributed Manufacturing	*Microproduction Everywhere. Social, Local, Open and Connected Manufacturing* Discusses the development prospects and case studies of distributed micro-production from the perspective of social innovation.
Cindy Kohtala	2015	Distributed Manufacturing	*Addressing Sustainability in Research on Distributed Production: An Integrated Literature Review* In this literature review of distributed manufacturing, the discussion revolves around how distributed production extends the lifecycle of products and materials, how producers and consumers can establish close relationships through distributed production, and how to further explore the sustainable potential of distributed manufacturing.
Fox Stephen	2015	Distributed Manufacturing	*Moveable Factories: How to Enable Sustainable Widespread Manufacturing by Local People in Regions without Manufacturing Skills and Infrastructure* Discusses suitable products for distributed production and strategies for overcoming challenges encountered in the manufacturing process.
Ezio Manzini, Mugendi K. M'Rithaa	2016	Distributed Systems in Social innovation	*Distributed Systems and Cosmopolitan Localism: An Emerging Design Scenario for Resilient Societies* Explores the role and impact of distributed systems in social innovation.

(continued)

Table 11.3 (continued)

Researchers	Year	Field	Publication
Massimo Menichinelli	2016	Distributed Manufacturing	*Sustainable Production in Emerging Markets through Distributed Manufacturing Systems (DMS)* Discusses the sustainable potential of distributed manufacturing in emerging market contexts and proposes a research framework.
Jagjit Singh Sraia, Mukesh Kumara, Gary Graham, et al.	2016	Distributed Manufacturing	*Distributed Manufacturing: Scope, Challenges and Opportunities* Discusses the challenges of and opportunities distributed manufacturing provides.
Silvia, Emili, Fabrizio Ceschin, David Harrison	2016	Distributed Energy Generation	*Product–Service System Applied to Distributed Renewable Energy: A Classification System, 15 Archetypal Models and a Strategic Design Tool* How to apply strategies and tool development for product-service system design to distributed energy.
Emili Silvia	2017	Distributed Energy Generation	*Designing Product-Service Systems Applied to Distributed Renewable Energy in Low-Income and Developing Contexts: A Strategic Design Toolkit* LeNSin project supports doctoral research exploring applications of distributed energy and sustainable product-service system design in low-income areas, providing a toolkit of creative generation devices to assist businesses, designers, and stakeholders.
Aine Petrulaityte	2019	Distributed Manufacturing	*Distributed Manufacturing Applied to Product-Service System: A Scenario-based Design Toolkit* The LeNSin project supports doctoral research in designing product-service systems in distributed manufacturing, providing a toolkit for addressing upcoming scenarios and offering solutions.
Jairo da Costa Junior	2020	Distributed Energy Generation	*A Systems Design Approach to Sustainable Development—Embracing the Complexity of Energy Challenges in Low-income Markets* Delft University of Technology and a joint doctoral research project, supported by Brazil's National Council for Scientific and Technological Development, discussed utilizing systems thinking and distributed technology for sustainable product-service system design, aiming to address complexity in low-income energy markets.

Social Dimension: The social technology systems that a distributed economy fosters exhibit greater resilience and recoverability[6] than a centralized system, making them more vibrant. This model facilitates equitable access to the means of production and resources, helping to prevent the maintenance and emergence of monopolies. It encourages public participation in social innovation, maximizing the use of social capital and leveraging collective wisdom. A distributed economy also increases the number of local employment opportunities and promotes regional social development and prosperity and suits people's differentiated, personalized needs. Furthermore, the open, flat organizational structures inherent in a distributed economy promote collaboration among individuals and organizations, fostering democratic and equitable decision-making processes. These structures also facilitate the flow and sharing of resources, information, and knowledge.

Economic Dimension: A distributed economy has the potential to create more job opportunities than a centralized economy, while stimulating local economic development. Small-scale, localized economic organizations within this model display increased flexibility and adaptability to market risks, significantly reducing distribution and transportation costs while minimizing delivery time and loss. This enhances market competitiveness. Although a centralized economy benefits from the efficiency of large-scale production, a distributed one offers a greater number of customized products and services to meet individual needs, which confers additional quality characteristics and advantages. This model may also establish a new definition of success.

11.4 Case Studies of the Distributed Economy

The sustainability characteristics of a distributed economy are not merely theoretical, but also actively demonstrated in various projects. The following table describes cases according to LeNSin's seven categorizations (Table 11.4).

Distributed Renewable Energy: Ocean Sun Floating Solar Power Station
Distributed energy systems predominantly rely on clean energy sources such as solar, wind, bioenergy, hydro, and geothermal. These systems use localized, small-scale production units to provide clean, accessible energy, optimizing the use of local renewable resources. This approach reduces unnecessary infrastructure construction costs and minimizes environmental impacts, while improving quality of life for local residents. Additionally, distributed energy creates diverse options for energy users, making the system flexible and resilient to potential failures.

One notable example is from the Norwegian startup Ocean Sun. This distributed energy system offers solar power on water for coastal and riverside communities, such as fishing villages and small islands. It provides comprehensive services including system planning, design, installation, and maintenance. The system also

[6] *Design, when everybody designs: An introduction to design for social innovation*: 20–21.

Table 11.4 Case studies in a distributed economy model

Project	Type	Services Provided	Time	Country/ Region
Ocean Sun	Distributed Renewable Energy	Large-scale water area solar power supply	From 2018 to present	Norway/ Southeast Asia
Warka Water Tower	Distributed Water Systems	Rainwater collection facilities and support services	From 2016 to present	Africa
3D print valve for ventilators	Distributed Manufacturing	3D printing services and related technical support	From 2020 to present	Italy
Cloud Farming Tribe	Distributed Food Production	Rooftop planting and related support services	From 2016 to present	Hong Kong, China
Himalaya FM	Distributed Information/ Knowledge	Online radio platform	From 2013 to present	Mainland China
Linux	Distributed Software	Open-source software	From 1991 to present	United States
Stable Diffusion	Distributed Design	Distributed online design creative platform	From 2021 to present	United States

Fig. 11.5 Offshore Floating Solar Power Station (Image source: https://oceansun.no)

supports the connection of multiple, parallel, small floating power stations, which provide power for large facilities, such as offshore seawater desalination plants (Fig. 11.5).

Distributed Water System: Warka Water Tower

A distributed water management system establishes localized, small-scale water collection or treatment units, including rainwater collection, sewage treatment, and seawater desalination systems. Through professional planning and design, these systems promote collaboration among local organizations and individuals, significantly reducing the construction and operation costs of facilities and creating resilient, adaptable local water supply solutions.

A representative case of a distributed water system is the Warka Water Tower, built in Ethiopia by Italian architect Arturo Vittori. Vittori designed a simple device for collecting rain and fog, for local residents (Fig. 11.6). Constructed from readily available bamboo, the Warka Water Tower is an open-source project that meets basic water collection needs. Building on this success, the designer further conceptualized the Warka Water Village, integrating food production and energy networks related to water with the Warka Water Tower. After receiving basic training, local residents can construct their own Warka Water Towers, tailoring the design to meet the specific conditions of their region.

Distributed Manufacturing: 3D Printed Respiratory Valves

The purpose of distributed manufacturing is to establish localized small-scale production units, create customized products, and meet individual needs, while reducing the costs of transporting materials during production and encouraging localized cooperative relationships. The combination of small, efficient production equipment and social networks generates opportunities for local consumption.

Fig. 11.6 Warka water tower (Image source: https://www.warkawater.org)

In 2020, at the beginning of the outbreak of COVID-19, the number of hospital admissions in Italy surged, causing over 2,500 deaths. The pandemic created a severe shortage of ventilator valves. In the span of six hours, Cristian Fracassi and Alessandro Ramaioli, designers at the startup Isinnova, working in collaboration with local Fablab founder Massimo Temporelli, designed and manufactured a simple valve for ICU patients in local hospitals. They continued to manufacture and donate the required valves until supply was restored.

Distributed Food Production: Rooftop Republic in Hong Kong

Distributed food production uses localized small-scale production units like rooftop and vertical farming to combine agricultural production with food consumption in specific localities. While enhancing local self-sufficiency, distributed food production ensures the freshness of food and enhances the food experience. At the same time, small-scale food production units can link to other food production units to form a distributed food network or connect with water or energy networks to further promote local sustainable development.

Rooftop Republic in Hong Kong is a cross-disciplinary team of architects, engineers, farmers, chefs, and nutritionists. They help users transform idle spaces into urban farms and organize workshops to provide customers with technical guidance on growing food and other plants, providing diversified urban agriculture solutions.

Distributed Information/Knowledge: Himalaya FM

In a distributed information/knowledge system, any actor can act as a unit (node) that produces knowledge and information, making dissemination flexible and efficient. This model represents a shift from a hierarchical structure to a flat, interconnected one, which evens the playing field and facilitates the acquisition of information and knowledge. For example, using internet platforms such as Himalaya FM, Twitch, TikTok, Facebook, YouTube, and Wikipedia, actors can establish independent information publishing platforms, share knowledge and news, and exchange information, interacting with other users in diverse ways (Fig. 11.7).

Distributed Design: LKKER Platform

As an emerging design development model, distributed design encourages collaboration and co-creation. It not only benefits the development of local, small, and medium-sized enterprises by providing new job opportunities, but its fully open project development concept also lowers the industry threshold, providing users with economical, efficient, fast solutions. Stable Diffusion is a deep, learning-based, text-to-image generation model that startup companies StabilityAI, CompVis, and Runway developed in collaboration in 2022. Unlike DALL-E and MidJourney, Stable Diffusion provides users with source code and pre-trained weights. Users own the copyrights to the images they generate and can freely use them for other commercial purposes. Additionally, to correct limitations in initial training, SD's end users can also undergo additional training, using fine-tuning to output images that are in line with their needs and personal style. A significant number of open-source communities have emerged around the creation of Stable Diffusion, such as Civitai, to which users can freely upload and share targeted image generation models, style plugins, and their own creative work.

Fig. 11.7 Wikipedia (Image source: foundation.wikimedia.org)

Fig. 11.8 OpenHarmony Open-Source Operating Systems (Image source: OpenAtom Foundation)

Distributed Software: Linux

Distributed software has fundamentally redefined the relationship between users and developers, effectively merging the two roles. This approach has resulted in a complete open-source model, beginning with the technology's underlying architecture. The high flexibility and technical effectiveness of distributed software allow any individual or institution to freely use the source code, modify it, and republish it, creating software tailored to specific needs. Linux exemplifies this model, not only as a distributed software system, but also as a landmark in the development of free software and open-source code. Another significant example is Huawei's OpenHarmony system, a distributed software developed from Huawei's Hongmeng System and currently managed by the OpenAtom Foundation (Fig. 11.8).

11.5 Potentials and Limitations

The evolution from decentralized to centralized, back to decentralized, and now to distributed systems reveals a continuous intertwining of technological innovation and social change, which leads to the emergence of new economic models. In the face of today's global social, environmental, and economic challenges, a distributed economy—with its focus on localized, decentralized, open, small-scale production units and numerous sustainable features—makes it one of the most promising economic models for the future.

However, rather than the total abolishment of large-scale production, our argument concentrates on finding a renewed balance between large- and small-scale and between resource flows that take place within and across regional boundaries.[7] In its current state of development, a distributed economy is a complement, rather than a complete alternative to today's mainstream economic model.

Design research and practice require grounding in a deep understanding of broad socioeconomic contexts. Distributed economies represent a significant development trend that has numerous advantages and potential for promoting sustainability. However, they are not universally applicable. Designers must carefully choose their methods and strategies based on specific contexts, since the distributed model is not an end in itself. Moreover, the complexity of the social and technical systems involved in these transformations indicates that traditional disciplines like product design, interaction design, and landscape design are no longer sufficient means for meaningful intervention. Only through interdisciplinary, integrated approaches that combine product, service, and system design can we ensure the sustainable transformation of the socio-economy.

---

[7] Johansson A, Kisch P, Mirata M. Distributed Economies—A New Engine For Innovation[J]. *Journal Of Cleaner Production*, 2005(13): 971–979.

Chapter 12
Systemic Design

Abstract In the real world, the complexity of design problems often surpasses our expectations. The various elements of these wicked problems are deeply intertwined, influencing one another and collectively creating systemic issues. Traditional design approaches frequently prove inefficient and inadequate to address the challenges we face.

Keywords System thinking · System science · Systemic design

The strategies discussed in previous chapters—green design, ecological design, biomimicry, product-service system design, social innovation design, and distributed system design—are all interrelated and sometimes mutually inclusive. Despite their focus on sustainable design, they are too small-scale and cannot address the complexity of real-world challenges. Their interconnection creates systemic problems, much like urban traffic issues, which involve urban planning, road management, transportation methods, energy technology, public awareness of regulations, and even regional environment and cultural traditions. That is why sustainable solutions for complex systems must emerge from systemic design. While the theory of systemic design is still evolving, a variety of practices at different scales and of various types offer valuable insights and inspiration.

12.1 The Meaning of Systemic Design

Contemporary society faces significant global changes that will profoundly impact future generations in terms of climate, environment, society, and the economy. Maintaining an isolated, mechanical, reductionist approach offers little hope for resolving these challenges. Indeed, this very mindset has contributed to their emergence. Addressing these issues requires a comprehensive, holistic approach, one which systemic design offers. The goal of systemic design extends beyond merely reducing the environmental impact of industrial products; systemic design aims to

X. Liu et al., *Design for Sustainability*, Sustainability and Interdisciplinarity,
https://doi.org/10.1007/978-981-96-8657-5_12

create sustainable production and consumption systems while fostering the sustainable development of local social, economic, and cultural contexts.

Systemic design focuses on behavior as a critical component for activating new relationships between various actors and managing resources effectively—transforming the output of one system into the input of another.[1] The production processes that comprise systemic design are not limited to industrial activities, but also encompass agricultural practices. Furthermore, systemic design involves the creation of local enterprises and open relationship networks, which can significantly contribute to regional economic revitalization. Thus, systemic designers do not only engage in creating appealing, ecofriendly products that meet market demands, but also develop sustainable services, economic and production systems, and educational experiences.

Achieving these objectives requires that systemic designers possess a broad range of skills and knowledge. They must master practical design techniques, possess a deep understanding of various design theories, and comprehend the complex interactions between disciplines such as economics, philosophy, politics, sociology, chemistry, biology, architecture, and engineering. For traditional designers, this task may seem overwhelming and excessively ambitious. However, systemic design requires interdisciplinary professional collaboration, clear delineation of boundaries, and establishment of new connections.[2] As Professor Bistagnino of the Politecnico di Torino observes: It is clear that systemic design does not have a sterile approach linked exclusively to technical-productive problems, because it embraces the values connected with "being" rather than "having," and changes evaluation and relational priorities in both productive and social realms.[3]

12.2 Thinking in Systems

12.2.1 Characteristics of Systems

We tend to follow Cartesian logic and associate every effect with a specific cause, a linear mode of thinking that oversimplifies reality. Generally, people find it difficult to grasp nonlinear phenomena, which are characteristic of natural and complex systems. Adopting a systemic approach therefore benefits us, not for the sake of complicating matters, but in order to acquire something resembling a holistic perspective of situations by understanding causal relationships that are not obvious.

---

[1] Bistagnino, L., *Micromacro: The Whole Of Micro Systemic Relations Generates The New Economic— Productive Model*[M]. Milano: Edizioni Ambiente, 2016: 21.

[2] Peruccio, P. P., Vrenna, M., Menzardi, P., et al. From "The Limits To Growth" To Systemic Design: Envisioning A Sustainable Future[C]// Proceedings of The 2018 Cumulus Conference: Diffused Transition And Design Opportunities: 751–759.

[3] Bistagnino, L. (2011). *Systemic design. Designing the productive and environmental sustainability* (2nd ed.). Bra, Italy: Slow Food Editore.

To do so, we must first define what constitutes a system. The term system comes from the Greek word *sýstēma*, meaning a gathering or complex. Jay W. Forrester, a pioneer in system dynamics and a professor at the MIT Sloan School of Management, defined a "system" means a grouping of parts that operate together for a common purpose[4] Similarly, the renowned environmental scientist Donella Meadows described a system as "a set of things …interconnected in such a way that they produce their own pattern of behavior over time."[5] According to Meadows, the components of a system include elements, interconnections, and functions, with complex systems typically comprising multiple subsystems.

Systems can be either open or closed. This principle applies to both natural and artificial systems. Cells, animals, and plants are open systems that require metabolization to sustain life. These systems obtain energy from external sources and maintain themselves as stable entities through material exchange, reflecting what we call a dissipative structure. Forests are also open systems that contain subsystems of animals and plants. Schools, regions, cities, and societies are likewise open systems that are artificial and dynamic. Social systems are complex and multifaceted, leading Qian Xuesen to define them as "open, complex, giant systems."[6] By contrast, a mechanical watch is a closed system,[7] comprising multiple elements that operate without material exchange, thus largely unaffected by external influences. However, a light bulb is an open system because it typically requires connection to the power grid (another system) to function.

To paraphrase Aristotle, systems are greater than the sum of their parts. We can define "A system is more than the sum of its parts. It may exhibit adaptive, dynamic, goal-seeking, self-preserving, and sometimes evolutionary behavior"[8] Compared to linear models, system models are more resilient, particularly open systems, which often possess the ability to self-organize and self-repair.

---

[4] Forrester, J. W. *Principles Of Systems*[M]. Cambridge, MA: Wright-Ellen Press, 1968: 1.

[5] Meadows, D. H. *Thinking In Systems*[M]. White River Junction, VT: Chelsea Green Publishing, 2008: 2

[6] In 1990, Qian Xuesen proposed a new classification of systems, dividing them into simple systems, simple giant systems, complex giant systems, and special complex giant systems. Among them, the social system, as the most complex system, is also called the special complex giant system. These systems are all open, exchanging matter, energy, and information with their environments, so they are called open complex giant systems. See: Qian Xuesen, Yu Jingyuan, Dai Ruwei. A new scientific field: open complex giant systems and their methodology[J]. *Nature Magazine*, 1990(13), 1: 3–10.

[7] A system that exchanges energy but not matter with the environment is called a closed system. See: Zhu Chuanzheng, Chu Ying, Xu Haihan. *Physical Chemistry*[M].2nd Edition. Beijing: Science Press, 2008: 11.

[8] Meadows, D. H. *Thinking In Systems*[M]. White River Junction, VT: Chelsea Green Publishing. 2008: 12

12.2.2 From System View to System Science

Systems thinking is not new. We can trace its roots back to Ancient Greece, a time before disciplinary boundaries emerged and a holistic worldview prevailed. The philosopher Heraclitus proposed an organic understanding of the world, asserting that everything is in constant motion, change, and transformation, governed by a harmonious force. The phrase "everything flows, nothing stands still" captures this way of thinking. Similarly, Ancient Chinese thinkers identified gold, wood, water, fire, and earth as the fundamental elements of the universe, using this simple systemic view to guide both their understanding of the cosmos and their social practices.

In the seventeenth century, French philosopher, scientist, and mathematician René Descartes introduced analytical thinking, which involved breaking down complex phenomena into smaller parts to understand the whole through the characteristics of its fragments.[9] Descartes' thinking significantly influenced the development of modern science, shifting the perspective of the world toward a more mechanical view. However, following his reductionist approach led to a fragmented understanding of nature, which contrasts sharply with a systems approach that emphasizes context, holism, and ecology. Although Descartes' method has dominated the period during which modernization occurred and continues to hold sway in social and economic fields, its limitations have become increasingly evident.

Since the twentieth century, the limitations of the reductionist model have prompted a shift toward viewing the universe once again as a dynamic, indivisible whole, with greater emphasis on the relationships among its elements.[10] This system view represents an attitude rather than a precise theory; it underscores the importance of understanding the interconnectedness and wholeness of systems rather than treating them as isolated, closed entities.

The development of systems science began in the 1920s, gaining momentum in 1937 when Austrian biologist Ludwig von Bertalanffy introduced the concept of general systems theory at a philosophical discussion at the University of Chicago. Subsequently, following the introduction of operations research, systems engineering methods became widely applied in communication technology, military, and aviation. By the 1950s, universities had established theoretical research and teaching in systems science. *In 1954,* Bertalanffy and others founded the *Society for General Systems Theory*. Significant advances were made in the field, with H.H. Good and R.E. McWhirter's 1957 article, "Systems Engineering: An Introduction to the Design of Large-Scale Systems," and institutions like MIT offering courses in systems engineering.

By the 1960s, systems science had become widely recognized in the West and the Soviet Union. Bertalanffy published *General Systems Theory: Foundations,*

---

[9] Capra, F. *The Web Of Life: A New Scientific Understanding Of Living Systems*[M]. New York, NY: Doubleday, 1996: 19.

[10] Capra, F., Luisi, P. L. *The Systems View of Life: A Unifying Vision* [M]. Cambridge: Cambridge University Press, 2014.

Development, Applications in 1968, signifying the maturity of the discipline. Systems science, a field encompassing mathematics and natural sciences, studies and controls complex systems, laying the groundwork for emerging disciplines such as control theory.[11] A system, as Bertalanffy defined, is a set of elements in their own interrelationships and in interrelationships with the environment. Since the 1980s, the emergence of nonlinear science and complexity research has significantly contributed to the development of systems science. In the twenty-first century, systems science, as an interdisciplinary field, has become the frontier and focal point of international scientific research, particularly due to its emphasis on the study of complex systems and complexity.

12.3 The Scope and Evolution of Systemic Design

Systemic design refers to the process of applying theoretical methods from system science to create new systems that maximize the satisfaction of specific target requirements. The application of systemic design spans a vast array of fields, including space programs, military deployments, urban planning, architectural design, software development, and product innovation. In contemporary practice, the most prevalent application of system design is in computer hardware and software systems, encompassing the design of basic equipment, components, modules, interactions, and data organization—a use of systems theory in the field of computing.[12] Additionally, as discussed in Chap. 8, system-level design strategies inspired by nature are becoming increasingly significant. Ecological researchers drawing on a natural system view are conducting analogous design experiments. To enhance our understanding of Earth's ecosystem, scientists have begun constructing artificial ecosystems, such as the renowned Biosphere 2, which represents a broad application of systemic design.

Biosphere 2 was an experimental artificial ecosystem designed to demonstrate the feasibility of a closed system capable of sustaining human life in outer space, in preparation for potential future migration from Earth. In 1972, the Soviet Union began constructing BIOS-3, a closed biological experimental base, over 7 years. However, it was not a completely closed system. In 1984, American oil magnate Edward Bass funded the Biosphere 2 project with $200,000,000,[13] which involved eight volunteers (called biospherians), who participated in a 2-year closed experiment. Scientists aimed to replicate Earth's biosphere in a completely artificial environment and to maintain the stable operation of this ecosystem.

[11] Von Bertalanffy, L. *General system theory*[M]. New York, NY: George Braziller, 1968.

[12] For details, see https://en.wiktionary.org/wiki/systems_design

[13] See the official website of Biosphere 2 at Columbia University, HTTPS://BIOSPHERE2.ORG. The name Biosphere 2 originates from its model, Biosphere 1, which is the Earth.

Fig. 12.1 Biosphere 2 (Image source: https://www.arizona.edu)

After 7 years of design and construction by a team of leading scientists and engineers, Biosphere 2 was completed in 1991, rising out of the desert in Oracle, Arizona, north of Tucson. The structure covered an area of 13,000 square meters, with a total volume of about 180,000 cubic meters, and was eight stories high. The main building, a sealed steel-frame structure made of glass, became the largest closed system in history. It housed five major ecosystems: rainforest, ocean, desert, grassland, and swamp. To simulate Earth's ecological environment, some 4000 species of animals, plants, and microorganisms were introduced from Australia, Africa, South America, and North America. The facility also included approximately 2020 square meters of farmland for growing various crops to provide food (Fig. 12.1).

Biosphere 2 was a near-perfect systemic design project, composed of three main parts: an above-ground airtight glass enclosure, an underground technical area, and a living area. According to the original plan, Biosphere 2 was intended to operate entirely independently: generating energy from solar panels surrounding the site; the plants in the artificial ecosystem would provide oxygen; and the extensive farm the biospherians managed produced the system's food, including vegetables and meat. However, during the first year of the experiment, the biospherians experienced severe hunger due to a lack of food production experience, resulting in significant weight loss, which only improved in the second year. Biosphere 2 also encountered problems due to fluctuations in oxygen and carbon dioxide levels, which reached dangerous concentrations at one point, requiring the management team to introduce pure oxygen from outside. Rising concentrations of carbon dioxide increased the acidity of water in the artificial ocean, causing the death of numerous fish and coral plants; a large number of pollinating insects went extinct, impeding plant reproduction; and certain pioneer species in the tropical rainforest area grew rapidly, resulting in weak branches that easily collapsed. Furthermore, uncontrolled rainfall caused the artificial desert to transform into jungle and grassland. Of the 25 vertebrate species initially introduced into Biosphere 2, only six survived.

Due to the challenges of maintaining the internal environment and ensuring the health of the scientists, Biosphere 2 ended on September 26, 1993, after 2 years and 20 h of existence. Columbia subsequently took charge of the $200 million project, which repurposed it as an environmental research laboratory. Despite its failure, the project contributed invaluable knowledge and experience, which continue to inspire our imagination. Humanity's past, present, and future all exist within a system where everything is interconnected. Whether imitating natural ecosystems or transforming artificial technical systems, the concepts and methods that govern systemic design are essential, if we want to expand our future living space and construct something resembling a sustainable future. However, despite rapid technological advances, humanity currently lacks a deep understanding of and control over the delicate, critical balance of natural ecosystems. Our path learning from nature remains long.

Emerging during the period when large-scale production prevailed and methods focused on manufacturing single products, design did not holistically address the systemic problems of production and consumption. In the 1950s and 1960s, under the influence of figures like Tomás Maldonado and Abraham Moles, the renowned Ulm School of Design (*die Hochschule für Gestaltung*) in Germany introduced new disciplines, such as control theory, systems theory, information theory, semiotics, and human-machine engineering, into its curriculum. The goal was to develop a new type of design methodology, which signified the school's embrace of system culture.[14] The Ulm School of Design directly and indirectly influenced the future of design culture and research, with many internationally renowned designers and educators studying and teaching at Ulm. Research in systems design has continued to evolve worldwide since. In the twenty-first century, systems design methods incorporating concepts of environmental sustainability have gained prominence. Among the leaders in this field, the Politecnico di Torino (PoliTo) in Italy offers a comprehensive master's program that integrates sustainability with systems design methods, making it unique. The Oslo School of Architecture and Design in Norway (AHO)and the Ontario College of Art and Design in Canada (OCAD) also offer courses related to systems design and have international influence. These institutions have expanded beyond the conventional confines of systems design and systems engineering focused on creation to adopt a holistic systemic design perspective, which includes social elements, such that system participants co-construct boundaries and forms.[15] These institutions have all joined the non-profit, Systemic Design Association, which promotes emerging practices and research in systems design.

In China, institutions such as Tsinghua University, Hunan University, Jiangnan University, and Tongji University have also integrated sustainability thinking into systems design education. At the beginning of the twenty-first century, Professor

---

[14] Peruccio, P. P. *Systemic Design: A Historical Perspective*[M]; Barbero, S. *Systemic Design Method Guide For Policymaking: A Circular Europe On The Way*. Turin, Italy: Umberto Allemandi, 2017: 68–74.

[15] Jones, P. The Systemic Turn: Leverage For World Changing[J]. *She JI: The Journal Of Design, Economics, And Innovation*, 2017, 3(3): 157–163.

Liu Guanzhong at the Academy of Fine Arts at Tsinghua University proposed the theory of Design Affairlogy. This theory combines systems design thinking with traditional Chinese perspectives on planning and creation, creating a unique blend of Chinese and international perspectives. The concept of affairlogy focuses on the relationship between people and objects, where reason represents the essence and internal logic of the contradictions between them. Affairlogy helps deconstruct the constraints of linear thinking, providing a distinctive way of understanding and researching problems that allows practitioners to see essence through phenomena, and thus stimulating innovative systems design thinking.[16]

Designers widely recognize the importance of systems design. Its philosophy has become foundational knowledge, constituting an essential part of contemporary design research and education.

12.4 Principles and Methods of Systemic Design

Systems design draws principles and methods from various schools of systems theories and design thinking. When addressing complex systemic issues, design thinking can synthesize insights from related disciplines. Through collaborative exploration, sensemaking, and innovative form generation, it guides human-centered design processes and contributes to the development of complex, multi-system, multi-stakeholder services and plans that impact various aspects of society.[17]

Currently, systems design primarily focuses on localized production and consumption systems, emphasizing the participants, assets, and resources involved in local socioeconomic activities. The goal is to create synergistic connections between production processes, natural systems, and surrounding communities.[18] This approach encourages the design of cyclical systems for materials, energy, and information to minimize waste and convert outputs (waste) from one system element into inputs (raw materials) for another, thereby fostering new localized value chains. System model inputs include resources like energy, water, and production materials, while outputs are finished products and waste. Waste may be repurposed as a resource for another production process, increasing the size of the cyclical system (Fig. 12.2).

The foundational logic of systems design adheres to ecological principles. Barry Commoner's four laws of ecology,[19] mentioned above, are applicable to

[16] Liu Guanzhong. *Methodology of Practical Logic* [M], Shanghai: Shanghai People's Fine Arts Publishing House, 2019.

[17] Same as 13.

[18] Barbero, S, Fassio, F. Energy And Food Production With A Systemic Approach[J]. *Environmental Quality Management*, 2011, 21(2): 57–74.

[19] Barry Commoner (1917–2012), hailed as the father of the modern environmental movement. He was the first person to link nature, humans, and technology, analyzing the root causes of environmental crises from an ecological perspective, revealing that the root of environmental crises lies in

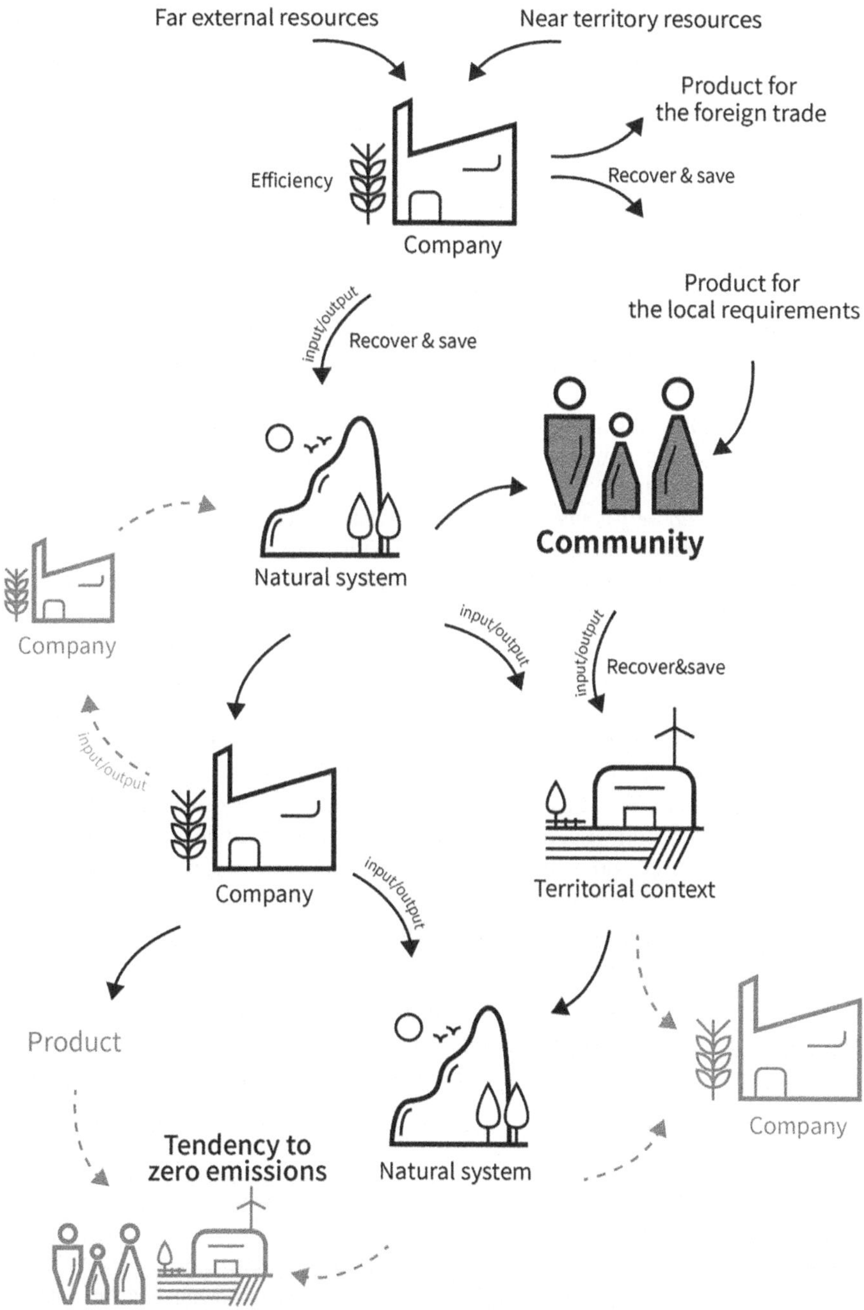

Fig. 12.2 Systemic productive model (Bistagnino, L. *Systemic Design: Designing The Productive And Environmental Sustainability* [M]. 2nd ED. Bra, Italy: Slow Food Editore, 2011) (Adapted from *Systemic design. Designing the productive and environmental sustainability* (p. 19), by L. Bistagnino, 2011, Bra, Italy: Slow Food Editore)

human-made systems. Based on these principles, Bistagnino[20] and the Zero Emissions Research Initiative (ZERI) have formulated the following idealized systems design guidelines:

1. *Output (waste) becomes input (resource):* There is no garbage or landfill in nature. We must reframe waste as a low-cost (or free) resource. Properly treated industrial or agricultural waste can serve as raw material for other production processes. The continuous flow of matter and energy across different systems can stimulate new production models, economic development, and job creation, while enabling system-wide circulation and interaction.
2. *Relationships create systems:* A system without relationships is merely a collection of non-functioning elements. In systems design, each element has equal importance and strategic significance. The relationships among individuals, industries, companies, and institutions within a system generate the energy that drives it, creating new dynamic interactions. These unique relationships make each system distinct.
3. *Systems should self-generate:* Ecosystems are dynamic and co-evolving, supporting each other and self-replicating. The ultimate goal in artificial systems is to achieve sustainability with elements benefitting each other, attainable only through cooperation rather than competition.
4. *Localize actions:* In a globalized world, it is crucial to expand, reproduce, and adapt localized actions. We must respect local interests and manage local social, cultural, and material resources wisely to foster social development. Taking such measures can mitigate the issues that may arise as a result of decentralized production methods and preserve local tangible and intangible cultural heritage.
5. *Focus on people:* Systems design focuses on human needs, while recognizing that humans are integrated into nature rather than separate from it. Systemic designers must consider the relationships that exist between communities and their territories,[21] artificial and natural environments, and humans and ecosystems. Thus, systems design requires an inclusive and interdisciplinary approach that accounts for diverse groups and numerous factors.

Designers should possess a foundational understanding of environmental sustainability, a sense of responsibility to all stakeholders, and the critical thinking skills to challenge engrained biases. Systems design also requires interdisciplinary

interactions between the human technological sphere and the natural ecological sphere, and proposing the famous Four Laws of Ecology. See: Krier, J. E. The Political Economy Of Barry Commoner [J]. *Environmental Law*, 1990, 20: 11–33.

[20] Same as 17.

[21] The concept of territory here originates from RAFFESTIN's (2012) proposal in the context of geopolitical geography. That is, territory is the result of a community projecting labor, in the biosociological category as a combination of energy and information, onto a given space. It superimposes human-related factors onto physical space, forming a complex relationship system. For details, see Raffestin, C. Space, Territory, and territoriality[J]. *Environment and planning D: Society and space*, 2012, 30(1): 121–141.

collaboration that makes room for diverse perspectives and improves efficiency and outcomes. Although the systems design process appears straightforward, it requires adapting to, sometimes unfamiliar, research plans and methodologies that emerge from specific project contexts. The process generally follows the following five steps:[22,23,24]

1. *Regional holistic research:* Conduct comprehensive research by reviewing previous publications and conducting field investigations to understand the project's background, current status, and uniqueness. This includes macro-scale data collection on the economy, society, and environment in a project area, as well as an in-depth analysis of specific industries, which focuses on material, energy, and capital flows. Although not all the data collected may be directly applicable to design, detailed analysis is imperative to developing a cogent project. Visualizations, such as system diagrams and infographics, can help analyze, diagnose, and communicate complex processes and relationships.
2. Best practice analysis: Learn from established concepts, methods, and technologies by analyzing best practices that address similar challenges. The goal is to identify common knowledge and determine how to adapt elements of prior practices. While systems design practices are often localized, making exact replication difficult, previous cases that have met with success provide valuable inspiration. Visual system diagrams are useful for conceiving these practices.
3. System problem identification: Use the visualized system map created in the first step, combined with insights from published resources and expert opinions, to diagnose and identify problems in existing systems, as well as specific production processes. These problems may include environmental issues (e.g., waste flows, carbon emissions generated by cross-regional logistics, undeveloped localized energy, waste of water resources) and social ones (e.g., loss of cultural heritage, rural depopulation, aging populations, left-behind children). Identifying these problems is critical, as it marks the starting point for genuinely engaging in systems design.
4. *Solution development:* This stage entails a series of iterative creative, development, and verification processes, which include:

- *From problem to opportunity*: Emphasize exploring the circular paths of output flows (for example, converting a waste stream from one production process into a resource for another production process) and input flows (for example, converting non-local resources into local resources).

[22] Battistoni, C., Barbero, S. Systemic Design, From The Content to the Structure of Education: New Educational Model[J]. *The Design Journal*, 2017,20(SUP1):S1336-S1354.

[23] Bistagnino, L. *Systemic Design: Methodology And Principles*[M]// Barbero, S. *Systemic Design Method Guide For Policymaking: A Circular Europe On The Way*. Turin, Italy: Umberto Allemandi, 2017: 75–82.

[24] Ceschin, F., Gaziulusoy, I. *Design For Sustainability: A Multi-Level Framework From Products To Socio-Technical Systems*[M]. London: Routledge, 2020.

- *Define a new system model*: Integrate design opportunities and paths into a new regional macro-system model, ensuring interconnected processes. Consider the complex relationships between input and output flows, the interventions of socioeconomic participants, and their potential interactions. This macro-system should represent an ideal, balanced, and harmonious relationship between regional production, community, and nature without hierarchy.[25]
- *Define output results:* Conduct preliminary evaluations and simulations to assess whether the new system model positively affects the region of implementation and examine any potential side effects such as impacts on the environment, economy, society, and culture. Effective evaluation requires external expert assessments to analyze the system's inputs, outputs, potential socioeconomic effects, and reactions from members of the public.
- *Solution implementation:* After preliminary verification of the output of the previous stages, implement the project stages. Real data collected during implementation can validate a design scheme's potential advantages and obstacles, facilitating subsequent adjustments. Through iterative cycles, the tangible improvements and long-term benefits of systems thinking become evident, gradually outlining a new, sustainable model for social, cultural, production, and economic activities.

12.5 Case Studies of Systemic Design

Below, we describe four distinct systemic design cases: the Lettuce House—Sustainable Living Laboratory of the Ecological Design Team at the Academy of Arts & Design at Tsinghua University; the coffee grounds recycling system project in collaboration with the Politecnico di Torino; MonViso Institute's Alpine Village Regeneration Project; and the Royal Seaport Ecological Community Project in Stockholm, Sweden. Although these projects differ in scale, scope, and design processes, they all share a foundation in sustainability objectives and systemic design methods. They employ innovative approaches to products, spaces, planning, services, and forms of organization to foster new production and consumption relationships. Due to the difficulty and complexity of implementing such projects and many clients' lack of understanding of sustainability, only a few design agencies provide comprehensive systemic design services. However, this trend is changing, and, in future, systemic design practices at different scales will become increasingly common.

Lettuce House: Sustainable Living Laboratory

Lettuce House is an experimental project the Ecological Design Team at the Academy of Arts & Design at Tsinghua University initiated in 2014. Unlike the systemic design approaches mentioned above, this project does not target a specific region or situation. Instead, it is a study and reflection of daily living and lifestyle, aimed at identifying

[25] Bistagnino, L. *Micromacro: The Whole of Micro Systemic Relations Generates The New Economic-Productive Model*[M]. Milano, Italy: Edizioni Ambiente, 2017.

problems and design opportunities. It integrates the concept of a circular economy with related technical methods to create a prototype for sustainable living.

A project partner resides in the Lettuce House, a living laboratory, maintaining subsystems and caring for the vegetables and flowers, but also testing the equipment's operation, calculating the typical energy consumption of typical families, and exploring opportunities for waste recycling and processing, while providing explanations and training for visitors.

The project consists of six 20-foot (six-meter) containers assembled into a small complex that includes bedrooms, living rooms (workrooms), dining rooms, kitchens, and toilets. There are also features such as roof gardens, outdoor platforms, rest areas, and other functional partitions. The roof garden not only produces organic vegetables and offers cool leisure space but also plays a role in building insulation.

The project's team focuses not only on the design and construction of container houses, but also on using clean energy, treating domestic waste, applying gray water facilities and biogas systems, and the development of home organic planting technology and equipment. The construction of a resource regeneration system is central to the project (Fig. 12.3). Wastewater from the shower, sink, washing machine,

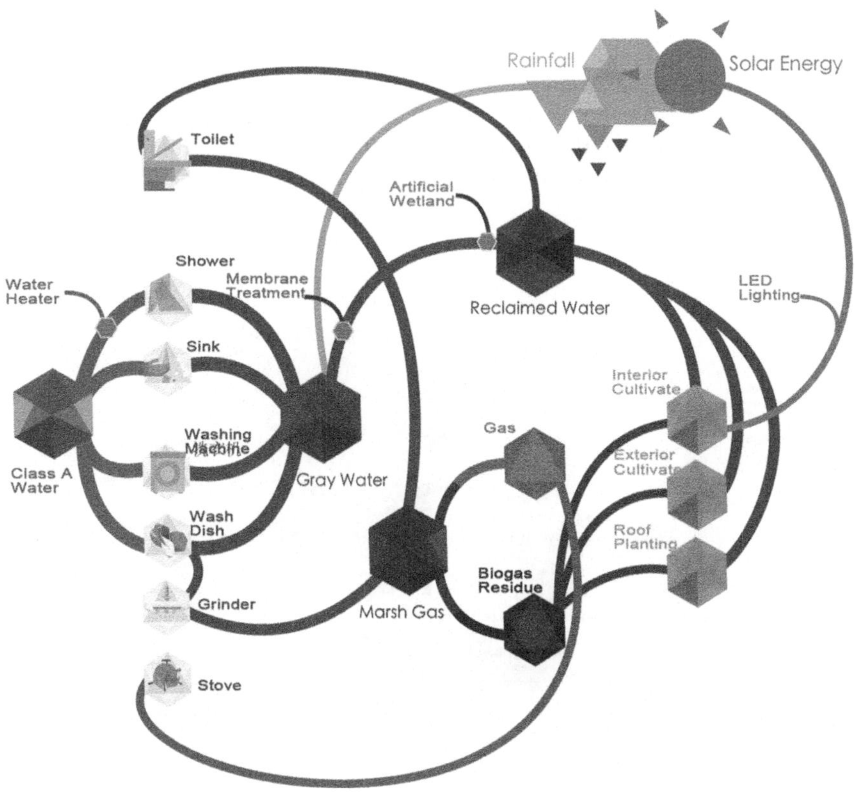

Fig. 12.3 Cycle system of Lettuce House (Image source: Eco-Design Institute, Academy of Arts & Design, Tsinghua University)

and vegetable washing pool, along with collected rainwater, is pooled into a gray water barrel. After simple purification, the water is stored in a middle water barrel for use in flushing toilets and irrigating roof gardens. Crushed kitchen waste, pooled with toilet waste, undergoes processing in a biogas barrel. Through sedimentation and anaerobic fermentation, the resulting biogas provides clean energy for the kitchen stove, while the biogas liquid and residue serve as organic fertilizer for the roof garden and indoor and outdoor plants. Additionally, solar energy supports lighting the indoor planting system. Despite some challenges with the biogas system, due to insufficient waste output, the project successfully achieved its systemic design goals.

The active participation and hands-on involvement by the project partner were critical to the design of this system. The final output of the project is not merely a residential system incorporating various ecological technologies, but also the practice and dissemination of a sustainable, people-centered lifestyle (Figs. 12.4 and 12.5). The project garnered widespread social attention and spanned imitations, with related information shared over 10 million times. In subsequent phases of the project, a number of professional institutions provided technical and equipment support, further creating buzz and encouraging public engagement in sustainable living practices.

Recycling Coffee Grounds

Coffee is integral to Italian culture, with high consumption levels across the country. A collaborative project between the Polytechnic University of Turin and Lavazza, one of Italy's largest coffee producers, explores reuse opportunities for coffee grounds. This project applies systems design principles and methods. It began with

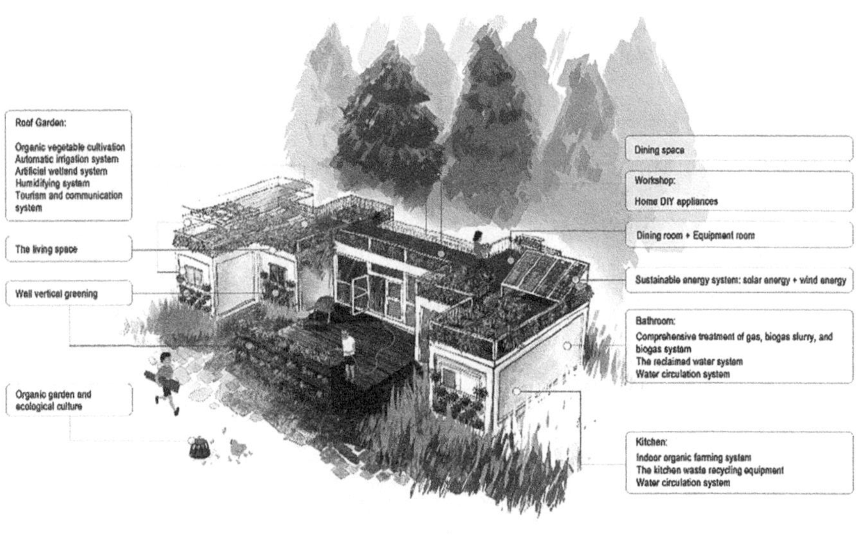

Fig. 12.4 Mockup of the Lettuce House (Image source: Eco-Design Institute, Academy of Arts & Design, Tsinghua University)

Fig. 12.5 Photos of the Lettuce House (Image source: Eco-Design Institute, Academy of Arts & Design, Tsinghua University)

comprehensive research into the regional economic and cultural characteristics. The focus then shifted to the material flow and technical analysis of each stage in the coffee industry chain. The research identifies a three-step process for reusing coffee grounds:

1. *Pharmaceutical raw materials:* Traditional caffeine extraction technology allows for extraction of lipids and waxes from discarded coffee grounds, which pharmaceutical companies can then use.
2. *Mushroom Cultivation:* The lipid extraction process produces a dense paste that serves as a substrate for cultivating nutritious and medicinal edible mushrooms.
3. *Fertilizer Production:* After mushroom cultivation, the used substrate has further applications in vermicomposting by cultivating worms.[26]

The project creators' primary focus is on designing processes and establishing cooperative relationships with stakeholders for mushroom cultivation using coffee grounds. The operational model involves collecting coffee grounds (outputs) from numerous bars and coffee shops and then, with support from local research laboratories, experimenting in mushroom production. Coffee grounds and other materials are used as substrates (inputs) in this process, which generates economic benefits (Figs. 12.6 and 12.7).

[26] Barbero. S., Toso, D. Systemic Design Of A Productive Chain: Reusing Coffee Waste As An Input To Agricultural Production[J]. *Environmental Quality Management*, 2010, 19(3): 67–77.

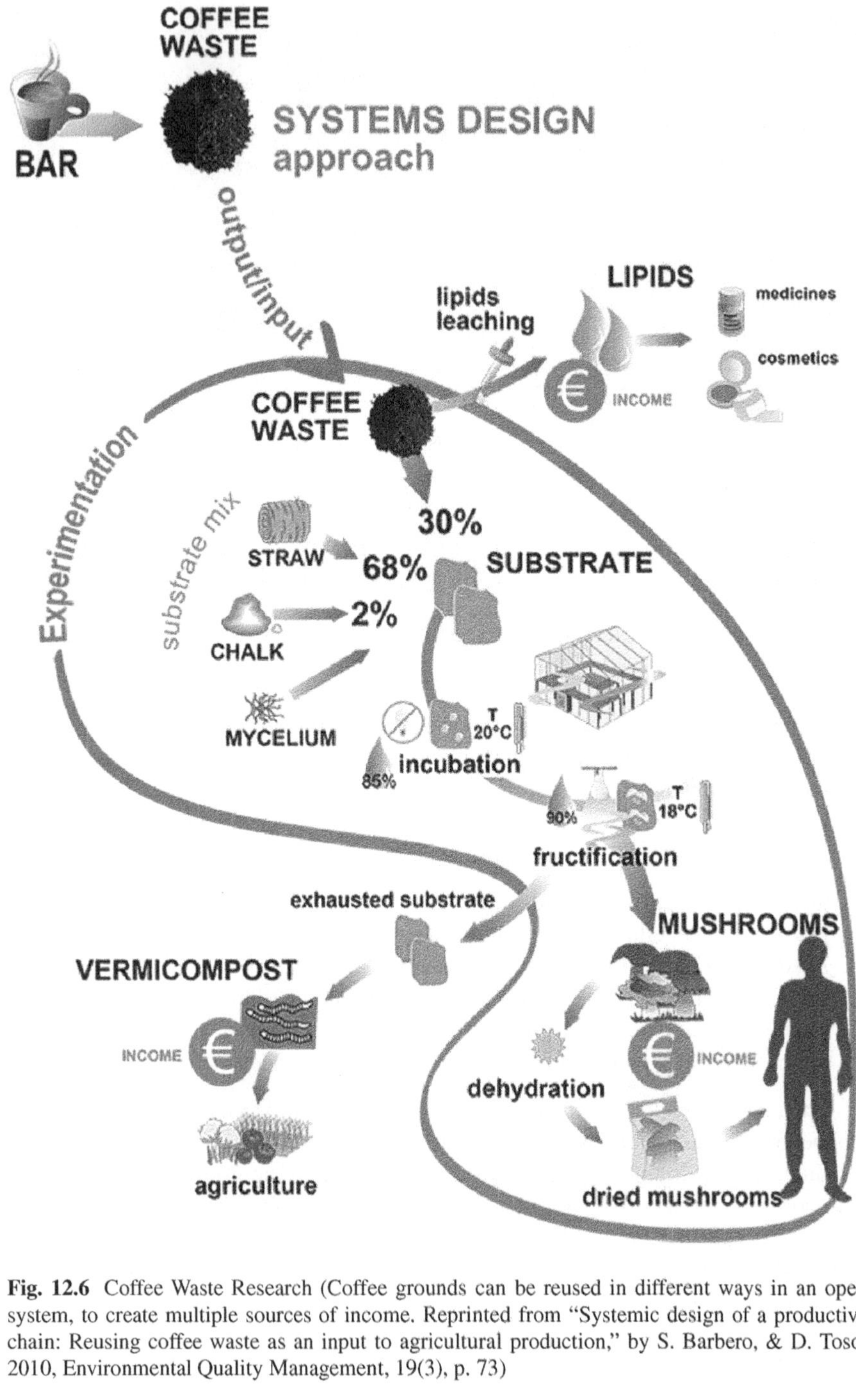

Fig. 12.6 Coffee Waste Research (Coffee grounds can be reused in different ways in an open system, to create multiple sources of income. Reprinted from "Systemic design of a productive chain: Reusing coffee waste as an input to agricultural production," by S. Barbero, & D. Toso, 2010, Environmental Quality Management, 19(3), p. 73)

Fig. 12.7 Mushroom cultivation using coffee grounds, straw, and ash as substrate (S. Barbero, D. Toso (2009)—Systemic Design, Buone previsioni dai fondi di caffè—Quaderni di Design, Time&Mind, Torino, 2008)

In 2015, under the guidance of the social enterprise Il Giardinone, 1500 kg of Lavazza coffee grounds helped produce 150 kg of high-quality oyster mushrooms with a protein content higher than mushrooms cultivated using standard methods. The project designers then repurposed the spent mushroom substrate as fertilizer to grow salad vegetables, resulting in a doubling of normal yields.[27] By 2016, Il Giardinone had launched the Fungo Box coffee ground mushroom cultivation kit, and in 2019, it introduced the Coffeefrom coffee cup made from coffee grounds.[28] This project reveals that we can use systems design to repurpose coffee grounds, typically viewed as waste, to create new value and new business models.

Regenerating Alpine Villages

Remote mountainous and inland rural areas worldwide often suffer from population loss and aging. Revitalizing these communities in crisis requires urgent implementation of targeted design interventions. MonViso, located in the province of Cuneo, Italy, is an innovative institution dedicated to sustainable regional development that tackles real-world social and environmental issues using interdisciplinary approaches. The institute employs designers and researchers from around the world working in various fields, all committed to enhancing public understanding of systems design and social transformation.[29]

The MonViso Institute uses unique systems design principles and methods. For their project aimed at revitalizing an alpine village, the researchers started by thoroughly examining the region's current conditions and historical background. They carefully analyzed the area's economic fluctuations, with a particular focus on its recovery process following the depopulation crisis.

This approach helped uncover the inherent reasons behind the region's economic resilience, such as the development and application of local products and materials, and the fusion of tradition with technology and social innovation. Specifically, the

[27] Lavazza. The Sustainability Report [R/OL]. [2021-08-20]. https://www.lavazzagroup.com/en/how-we-work/the-sustainability-report.html.

[28] See the company's website: https://www.ilgiardinone.it

[29] Luthe, T. Co-Designing A Real-World Laboratory for Systemic Design In The Italian Alps: How Complexity Shapes The Process[C/OL]//Proceedings Of Relating Systems Thinking And Design (RSD6) 2018 Symposium, Delft, The Netherlands, October, 2018. [2020-03-16]. https://rsdsymposium.org/co-designing-a-real-world-laboratory/.

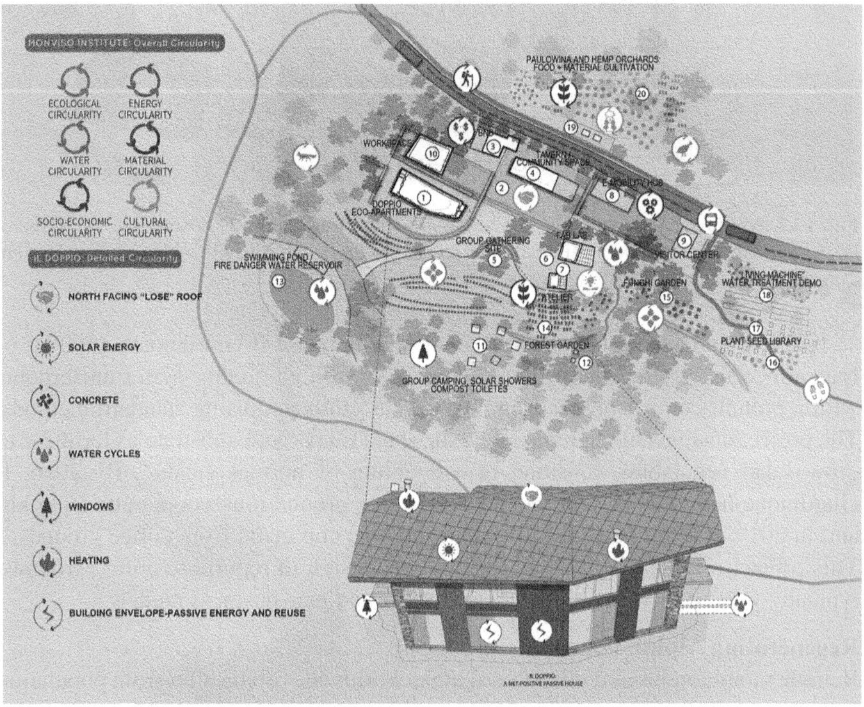

Fig. 12.8 The map of the MonViso Institute with the details of the systemic project. (Adapted from Reaching new heights, by H. Fitzpatrick, & T. Luthe, 2019, retrieved from https://www.systemsorienteddesign.net/index.php/397-eco-tourism)

project involves the partial or complete restoration of old farmhouses using traditional techniques, connecting the community to a self-sufficient system by designing water, energy, and transportation systems, and promoting production of industrial hemp[30] and innovative design of eco-products, such as snowboards made from local materials.[31] Figs. 12.8 and 12.9 present the project's overall cycle and local buildings renovated with green technology and recycled materials.

Given the inherent complexity of rural regeneration projects, the team recognized the critical value of social support networks and adopted an activism principle. This approach involved continuously seeking ways to optimize the system

[30] According to the unified standards set by the European Commission, industrial hemp refers to cannabis plants and their extracts with a tetrahydrocannabinol content of less than 0.3% (percentage of dry matter weight), called hemp in China. With modern technology, through research and development, we can extract raw materials from the roots, stems, leaves, skins, flowers, and seeds of industrial hemp, which are widely used in daily chemicals, materials, energy, medical, food, and oil substitution.

[31] See the academic poster: Reaching New Heights by Haley Fitzpatrick and Tobias Luthe at the Relating Systems Thinking And Design (RSD8) 2019 Symposium, available at https://www.systemsorienteddesign.net/index. php/397-eco-tourism.

Fig. 12.9 Aerial view of the MonViso Institute in Ostana (Old farmhouses are being rebuilt using traditional techniques and local materials. In the background Monviso, the highest mountain of the Cottian Alps, located in Italy close to the French border. Reprinted from Monviso Institute website, 2024, retrieved from http://monviso-institute.org Copyright 2024 by Tobias Luthe)

through co-designing and co-building with villagers. The project benefitted from active support and participation from the local government and residents. In addition to regular seminars, the designers invited students for short-term training and guided local people's engagement in the revival of their community. Although the project is ongoing, it is anticipated that systematic research and design will restore vitality and resilience to the declining village.

Royal Seaport Eco-Community in Sweden

Sweden is one of the most sustainable countries in the world, effectively making use of various renewable energy sources and maintaining very low carbon dioxide emissions. The Royal Seaport in Stockholm was once composed of port areas, industrial facilities, and hunting grounds. The designers seek to turn it into a world-class model of sustainable urban development. Initiated in 2008, the Royal Seaport project is expected to be fully functional by 2030. Upon completion, it will be Sweden's largest urban community, encompassing residential, office, leisure, and cultural spaces. In 2012, the first batch of residents moved into completed homes in the northern part of the project's territory.

As a model of urban systems design, the Royal Seaport project organizers conducted extensive foundational work before and during its implementation. This

Fig. 12.10 Panorama of Royal Seaport in Sweden (Image source: https://international.stockholm.se/city-development/the-royal-seaport/)

included technical preparation and experiments to develop the ecological cycle system, as well as organizing workshops and urban development dialogues with future community stakeholders, including residents, staff, government workers, researchers, municipal department representatives, and social organization staff members. The project leaders also established several special groups to concretize the target vision and developed a robust system of project supervision, evaluation, and feedback to manage, guide, and ensure the high quality of work methods (Fig. 12.10).

The project achieves sustainable urban development via four key design strategies: mixed-use urban programs, sustainable, convenient urban transportation, recycling and efficient resource use, and an environmentally friendly blue-green system.[32] Among these, the ecological circulation systemic design of energy, materials, and water, which serves business, industry, and daily life, is particularly noteworthy. This system includes eco-concept buildings, waste-recycling systems, infrastructure reuse, rainwater collection systems, smart grids, intelligent logistics hubs, and self-sufficient energy systems (Fig. 12.11).[33]

The Hammarby Sjöstad eco-city, the planning and construction of which began in the 1990s, already integrates the flow of energy, water, and materials into an overall system, which connects it as an open system to the city of Stockholm.[34] This nature-based ecological cycle model not only inspired the Royal Seaport project, but also provided crucial technical support for its implementation. In 2010,

[32] Blue-green systems refer to the blue-green infrastructure in cities that can produce resources (raw materials, food, clean air, and water) and absorb the waste urban activities generate. Green infrastructure is related to vegetation, such as green roofs, green walls, mini parks, urban farms, and forests, while blue infrastructure pertains to water, including permeable surfaces, reed beds, reservoirs, etc. See footnote 2 for details.

[33] Williams, J. The Circular Regeneration Of A Seaport[J]. *Sustainability*, 2019, 11(12): 1–27.

[34] Ranhagen, U., Frostell B. Eco-Cycle Model 2.0 For Stockholm Royal Seaport City District: Feasibility Study [R/OL]. KTH School Of Architecture And The Built Environment, 2014. [2021-08-20]. https://www.diva-portal.org/smash/get/diva2:736415/fulltext01.pdf

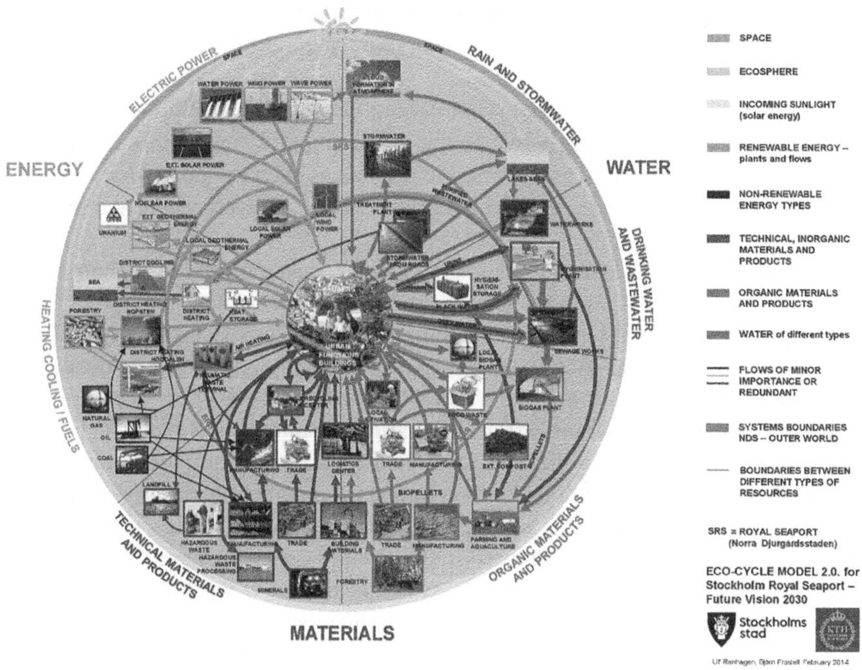

Fig. 12.11 Ecological system model of Stockholm's Royal Seaport district (Image source: http://www.diva-portal.org)

then-Vice President Xi Jinping made a special visit to the Hammarby eco-city,[35] calling attention to its pioneering design concepts and its potential value to China's ecological civilization and future urban development.

12.6 Systemic Design in Definition

Systemic thinking originates from human efforts to learn from and imitate natural ecosystems. While advances in contemporary science and technology, particularly in artificial intelligence and biotechnology, have significantly promoted the development of system-related disciplines, humankind has yet to fully comprehend nature in terms of the intricate knowledge and inherent connections of all elements in it as a complex, open system. Consequently, the design of artificial technological systems remains at a rudimentary, immature stage. In contrast, human social systems, which have evolved over hundreds of thousands of years, exhibit a complexity comparable to natural systems. The initiative, creativity, potential uncertainty, and

[35] Xinhua News Agency. Xi Jinping visits Hammarby Eco-city[EB/OL]. (2010-03-30) [2021-08-22]. http://www.gov.cn/govweb/jrzg/2010-03/30/content_1569457.htm

destructive power of people—one of the system's elements—render understanding, predicting, and, most importantly, changing relationships in social systems extremely challenging.

The purpose of systemic design is to equip designers, broadly defined, so that they may confront and address sustainable development issues within comprehensive social, environmental, and economic systems. Given the complexity of these challenges and the intricate, intertwined relationships involved, systems thinking is essential.

One advantage professional designers enjoy in systemic design is their exceptional ability to express complex ideas visually. Whether relating the intricate relationships of existing systems or developing system solutions, visual expression is critical for conveying abstract concepts and relationships. In systemic design, we should not focus on creating visually dazzling effects, but rather on clarifying logical relationships and providing clear explanations. While the style of visual expression can vary depending on the context or theme, design performance should avoid a formulaic approach. Additionally, in systemic design, the role of the designer as a coordinator is paramount. The long-term nature of systemic design projects, coupled with the involvement of numerous stakeholders, may present cultural, social, and economic challenges. A qualified systemic designer, with a comprehensive grasp of system goals and overall planning, an understanding of human nature and needs, and expertise in design strategies and methods, thus plays a crucial role in coordinating relationships among various elements and actors and ensuring project progress.

The theory of systemic design is still evolving, with scholars differing in their understandings and interpretations of what constitutes a system. The concept of systemic design is likewise in flux. Moreover, practical applications have largely been confined to specific contexts. Due to regional differences, we cannot easily replicate or generalize the results of practical and theoretical initiatives. Therefore, making progress in systemic design in future will require a broadening of the diversity of design experiments to further refine and improve theories and methods.

Current systemic design research tends to emphasize the construction of technical recycling systems and regional economic systems, often neglecting intervention in people's consumption and other habitual behaviors. This oversight may result in situations where the environmental benefits we acquire from system optimization are continuously offset by increasing total consumption. As we have repeatedly emphasized in this book, only by changing concepts and culture can we fundamentally reduce the negative impact of human development on the environment. This means that systemic design projects must not only adopt a combination of complementary strategies, but also attempt to cultivate the metaphorical soil of a sustainable society. Designers must adopt proactive stances and innovative working methods, collaborating closely with policymakers, managers, engineers, educators, and the public to establish new standards and trends, thereby fostering growth of sustainable culture.

Conclusion

This conclusion marks both the end of this book and the beginning of a new chapter. The future trajectory of the world, as the COVID-19 pandemic has made clear in particular, remains uncertain. However, the mission of design is to actively address these uncertainties and propose innovative solutions and visions for the future. In this era of significant change, design will inevitably develop new concepts, theories, and methods to confront emerging conflicts and problems, enriching itself and meeting these challenges. We will witness and participate in this transformative design practice together.

In the preceding twelve chapters, using theoretical discussions and case analyses, and with detailed introductions to various design strategies, we have aimed to provide readers with a roadmap. This roadmap clarifies the evolutionary path and development trends of sustainable design, encouraging researchers, practitioners, and learners in the design field to reflect on the role and value of design. It also aims to equip them with the willingness, knowledge, and ability to engage in the design practice necessary for sustainable social transformation.

While we have endeavored to cover the main areas of sustainable design, certain topics, though briefly addressed in earlier chapters, have not been explored in dedicated chapters. These topics include design promoting a circular economy, sustainable consumption and behavior design, transformation design, and sustainable design evaluation systems. We will address these topics in our future work, and we invite readers to await these developments and provide constructive feedback.

The following is a summary of the main points of this book:

First, sustainability is not a new issue. Since the time of clan societies, there has been a conflict between human survival and development and the environment, increasing with the spread of agriculture, the industrial revolution, and population growth. In recent decades, due to the prevalence of consumerist culture, the conflict between economic development and the environment has intensified. Issues such as resource allocation and social equity have come to the fore, making sustainable development of the environment, society, and economy a global focus and a challenge faced by all of humanity. We have moved from an empty to a crowded world.

X. Liu et al., *Design for Sustainability*, Sustainability and Interdisciplinarity, https://doi.org/10.1007/978-981-96-8657-5

We need a new Enlightenment to surmount these difficulties.[1] With climate change, environmental crises, technological alienation, and political disputes, sustainability may remain a topic of discussion over the long term. No single country can achieve sustainable development alone. As a community with a shared future, humankind's road to sustainable development will undoubtedly be long and arduous, but sharing a common vision and dream of sustainability brings with it profound meaning.

Second, sustainable development entails harmonious coexistence between humans and nature. Conceptually, sustainable development, whose practitioners seek to sustain the continuation and development of humanity, seems unrelated to nature. However, the planet does not care whether humans exist. We are merely a small part of the natural ecosystem. Although we consider ourselves superior, able to dominate all species and create great civilizations, we will ultimately find that nature is the foundation and premise for all creation and human existence. The arrogant attitude of anthropocentrism intensifies the conflict between humans and nature, inevitably leading to a crisis in human society's survival. On the other hand, environmental centrism misinterprets the core goal of sustainable development, ignoring the basic right to survival and development of people in underdeveloped areas. Humans and nature constitute a community of shared destiny, aligning with the traditional Chinese ideal of unity of man and nature. Therefore, achieving sustainable development requires revering and protecting nature as the mother of life and learning from nature's wisdom to live harmoniously with all things.

Third, sustainable design encompasses all design activities aiming to achieving sustainable development, including research, education, and practice. Design is no longer solely a business-oriented creative activity. Sustainable development is not only the research field of politicians, scientists, sociologists, and economists, but also the shared responsibility of humanity and a core issue for designers. The World Design Organization defines design as a response to social, economic, environmental, and ethical issues, aiming to create a better world.[2] The boundaries of the design field are expanding, and design researchers, educators, and practitioners bear increasing social responsibilities. Doing responsible design is not an empty slogan but rather constitutes an emerging consensus. Additionally, the design community should reflect on the innovation-only approach. Innovation has always been a keyword in design, but we need to consider why we innovate and the direction of innovation. Innovation is only a means. If pursued for its own sake, design will lose its core purpose and direction. Therefore, sustainability should become the basic evaluation standard for design innovation.

Fourth, sustainable design should integrate wisdom from both the East and West. Originally a foreign concept, sustainable design arose from reflecting on the unsustainable production and lifestyle of Western society. China, after decades of rapid

---

[1] Wei Bole, Vijayakumar. Flipping the Limit: The Awakening Road of Ecological Civilization [M]. Cheng Yiheng, trans. Shanghai: Tongji University Press, 2019: viii. Ernst Ulrich von Waizsacker, Anders Wijkman, Come On!: Capitalism, Short-termism, Population and the Destruction of the Planet.

[2] Official website of the World Design Organization: https://wdo.org/

development, faces similar, if not more complex, environmental and social problems. At this new stage of transformation, designers need to explore various development paths. The design community should draw on traditional Chinese ways of thinking and creation, such as that which expounds the unity of man and nature, making full use of resources, moderation, and the doctrine of the means. These concepts align with modern sustainability ideas and are holistic and systematic. Constructing a complete discourse and theoretical system that responds to contemporary problems is an urgent task. Simultaneously, we need to embrace contemporary theoretical achievements and render them efficacious. Overemphasizing the differences between the East and West brings no benefits. We must recognize that inclusiveness is part of the cultural genes of the Chinese nation. When facing common crises and challenges, the fusion of Eastern and Western wisdom can lead to wise choices.

Fifth, there is no absolute sustainability. There is no best, perfect, or absolute sustainable design solution. Life Cycle Assessment (LCA) remains the most effective method for assessing the sustainability of products and systems from an environmental perspective. Social sustainability assessment is however much more complex than LCA. Comprehensive assessments of environment, society, and economy involve multiple factors over long periods of time. Sustainability is a goal, direction, and vision that we pursue, with design concepts, strategies, methods, and standards that constantly evolve. As designers' understanding of sustainability develops and new materials and technologies emerge, what is consider good design today may become problematic and flawed tomorrow, necessitating further corrections and improvements. We must compare good sustainable design with similar and previous designs. Through constant comparison and improvement, we can establish a deep understanding of sustainable design.

Sixth, integrated system design solutions are necessary. The sustainable development issues we face are systemic problems involving multiple elements such as the economy, environment, society, ethics, and culture. Addressing these issues requires integrated system design solutions rather than simply technological improvements or product innovations. For example, designing an environmentally friendly electric car does not solve traffic congestion. Instead, designing a shared travel service platform, creating more bicycle routes, pedestrian paths, and urban landscapes can improve the transportation system. Systems thinking helps identify the roots of problems, integrate design strategies, and implement them step by step. System design can start with small projects like community gardens and canteens for the elderly. As John Thackara notes in his 2017 book, *How to Thrive in the Next Economy*, countless small actions eventually change the overall situation. Small-scale subversions accumulate until they cross an unknown threshold, changing the entire system.[3]

[3] Sakala. The Call of the New Economy: Designing Tomorrow's World [M]. Ma Jin, Ma Yue, trans. Shanghai: Tongji University Press, 2019: 270.

Seventh, sustainable design requires a transformation of design roles and methods. Sustainable design involves collaborating with experts from different fields and coordinating various interests to co-create with different groups. As sustainable design faces increasingly complex system problems and most outputs consist of comprehensive system solutions, traditional design roles and working methods will no longer be suitable. Design practitioners will need a macro perspective, systematic thinking, forward-looking ideas, imagination, and empathy. They must integrate themselves into design objects and situations, uncover real problems, understand real demands, and propose meaningful, valuable system solutions through interdisciplinary collaborative innovation. This change implies a transformation in the knowledge structures, abilities, and working methods of traditional designers. While creative ability, control of modeling terminology, and visual expression will remain core competencies, the ability to promote interpersonal communication and communication skills will also become crucial.

Eighth, sustainable design is a collective learning process. Sustainable design theories and methods are evolving. Due to the lack of clear evaluation standards and increasingly complex, open system problems, in practice, many misconceptions and challenges arise. We cannot expect a single leader to emerge to specify the path of development and provide standard answers. Although numerous designers in the past have conducted meaningful, forward-looking theoretical research and practices, while achieving remarkable results and providing important references, each context differs. When regional environment, economic conditions, social culture, institutions, situations, and populations change, we cannot simply replicate theoretical models and practical experiences because their effectiveness becomes questionable. Sustainable design may entail grand system changes or product and service innovations. It is a continuously developing research field and a process of collective learning. Only through continuous learning, experimentation, reflection, and communication can we achieve a unity of knowledge and action and develop sustainable design ideas and methods.

Fundamentally, design is a means, neither inherently noble nor humble. Design can be part of the solution to problems but it can also create them. Designers and developers have developed many remarkable products, beneficial to the world and humanity, but unfortunately, they have also contributed to many harmful products.[4] The impact of design—positive or negative—is uncertain and not immediately apparent. However, designers are not without choices. Designers play a pivotal role in understanding human nature, identifying needs, raising questions, coordinating communication, and creating solutions. We cannot underestimate the impact a design solution has on the world, the changes it effects with regard to the environment and the population. As a driver of contemporary cultural innovation, design may play a leadership role in the transition to a sustainable future. As China embarks on a new journey towards developing an ecological civilization, designers have

[4] Shedroff N. Design is The Solution: The Future of Design Will Be Sustainable [EB/OL]. [2021-08-10]. http://www.experiencedesignbooks.com. P258.

ample reason to make positive choices, creating kind, responsible designs for themselves, their communities, and their countries, to ensure the sustainable development of human society.

This book provides design strategies that are neither the only path we may take nor fixed doctrines. We hope these efforts will inspire readers and motivate everyone to explore new sustainable strategies and methods. We believe that design can truly change the world.

References

1. Hobsbawm, Eric. 1999. *Industry and empire: From 1750 to the present day[M]*. London: Penguin Group(CA).
2. Manzini, Ezio. 2015. *Design, when everybody designs: An introduction to design for social innovation[M]*. Cambridge, MA: The MIT Press.
3. Commoner, Barry. 2020. *The closing circle: Nature, man, and technology[M]*. New York: Dover Publications; Reprint edition.
4. Hawken, Paul, L. Amory Lovins, and Hunter Lovins. 2000. *Natural capitalism: Creating the next industrial revolution[M]*. New York: Back Bay.
5. Bhamra, T., D. Lilley, and T. Tang. 2011. Design for sustainable behaviour: Using products to change consumer behaviour[J]. *The Design Journal* 14 (4): 427–445.
6. Sevaldson, B. 2017. *Proceedings of relating systems thinking and design (Rsd6) 2017 symposium[C]*. Norway: Systemic Design Research Network (SDA).
7. Boehnert, J. 2018. *Design, ecology, politics: towards the ecocene[M]*. New York: Bloomsbury.
8. Boulding, K. 1966. The economics of the coming spaceship earth[M]. In *Environmental quality in a growing economy*, ed. H. Jarrett, 1–14. Baltimore, MD: Resources for the Future/Johns Hopkins University Press.
9. Brantlinger, P. 1996. A postindustrial prelude to postcolonialism: John Ruskin, William Morris, and Gandhism. *Critical Inquiry* 22 (3): 466–485.
10. Brezet, H., and C. Van Hemel. 1997. *Ecodesign: A promising approach to sustainable production and consumption[M]*. Paris: United Nations Environment Programme.
11. Brian Arthur, W. 2009. *Natural capitalism: Creating the next industrial revolution[M]*. New York: Free Press.
12. Caradonna, J. L. 2014. *Sustainability: A history[M]*. Oxford: Oxford University Press.
13. Chapman, J. 2017. *Routledge handbook of sustainable product design[M]*. Abingdon, Oxon: Routledge.
14. Coulouris, G., J. Dollimore, T. Kindberg, et al. 2011. *Distributed systems: Concepts and design[M]*. 5th ed. Boston: Addison-Wesley.
15. Joseph R. DesJardins. 2005. *Environmental ethics: An introduction to environmental philosophy[M]*. Calif: Wadsworth Publishing.
16. Jui-Che, Tu. 2002. *Product sustainable design: Green design in theory and practice[M]*. The Asia Pacific School of Business.
17. Ellen MacArthur Foundation. Reuse: rethinking packaging[R/OL]. 2019 [2021-06-05]. https://www.ellenmacarthurfoundation.org/assets/downloads/Reuse.pdf.
18. Dewberry E, Goggin P. Ecodesign & beyond: steps towards sustainability[EB/OL]. 1995 [2021-06-05]. http://oro.open.ac.uk/29316/.

X. Liu et al., *Design for Sustainability*, Sustainability and Interdisciplinarity,
https://doi.org/10.1007/978-981-96-8657-5

19. Manzini, E., and M. K. M'rithaa. 2016. Distributed system and cosmopolitan localism: An emerging design scenario for resilient societies[J]. *Sustainable Development* 24 (5): 275–280.
20. Spier, Fred. 2015. *Big history and the future of humanity[M]*. 2nd ed. Malden: Wiley-Blackwell.
21. Grunwald, Armin, and Rafaela Hillerbrand. 2021. *Handbuch technikethik[M]*. 2nd ed. Stuttgart: J.B. Metzler.
22. Daly, Herman E. 2016. *From uneconomic growth to steady-state economy[M]*. Cheltenham: Edward Elgar Publishing.
23. Daly, Herman E., and Joshua Farley. 2010. *Ecological economics principles and applications[M]*. 2nd ed. Washington, DC: Island Press.
24. International Institute for Industrial Environmental Economics. 2009. *The future is distributed: A vision of sustainable economies[R]*. Lund: Lund University.
25. Irwin T. The emerging transition design approach[M/OL]// Storni C, Leahy K, et al. Design as a catalyst for change-DRS International Conference 2018, 25–28 June, Limerick, Ireland. [2021-09-04]. doi:https://doi.org/10.21606/drs.2018.210.
26. Irwin, T. 2015. Transition design: A proposal for a new area of design practice, study, and research[J]. *Design and Culture* 7:229–246.
27. Jégou F, Manzini E. Collaborative service: social innovation and design for sustainability [M]. Milan: Edizioni POLI.design, 2008. [28] Jin G T. Systematic Philosophy[M]. Lujiang Publishing House, 2019.
28. Kelly, K. 1995. *Out of control: The new biology of machines, social systems, and the economic world[M]*. Basic Books.
29. Kallipoliti, L. 2018. *History of ecological design[M/OL]*. Oxford: Oxford University Press. https://doi.org/10.1093/acrefore/9780199389414.013.144.
30. Kossoff, G. 2016. Holism and the reconstitution of everyday life: A framework for transition to a sustainable society. *Design Philosophy Papers* 13 (1): 25–38.
31. Lewis, H., and J. Gertsakis. 2001. *Design + environment: a global guide to designing greener goods[M]*. Sheffield, UK: Greenleaf Publishing Limited.
32. Liu X, Vrenna M. Study on systemic design based on sustainability[J]. Chinese Journal of Design Zhuangshi, 2021(12):25–33.
33. Lou, Y. Q. 2016. Study on sustainable design: The tools, theory and tendency of design[J]. *Chinese Journal of Idea & Design* 5:22–25.
34. Nash, R. F. 1989. *The rights of nature: A history of environmental ethics[M]*. University of Wisconsin Press.
35. Malpass, M. 2017. *Critical Design in Context: History, theory, and practice[M]*. Bloomsbury Academic.
36. Maldonado, T. 1970. *Design, nature, and revolution: Toward a critical ecology[M]*. Minneapolis, MN: University of Minnesota Press.
37. Matson, P., K. Andersson, and W. C. Clark. 2016. *Pursuing sustainability: A guide to the science and practice[M]*. Princeton, NJ: Princeton University Press.
38. Meadows, D. H., D. L. Meadows, J. Randers, et al. 1972. *The limits to growth: A report for the Club of Rome's project on the predicament of mankind[M]*. New York, NY: Universe Books.
39. Mirata, M., H. Nilsson, and J. Kuisma. 2005. Production systems aligned with distributed economies: Examples from energy and biomass sectors. *Journal of Cleaner Production* 13 (10–11): 981–991.
40. Pickett, S. T. A., J. Kolasa, C. G. Jones, et al. 2010. *Ecological understanding: The nature of theory and the theory of nature[M]*. Academic Press.
41. Mokyr, J. 2018. *A culture of growth: The origins of the modern economy[M]*. Princeton University Press.
42. Rauch, E., P. Dallasega, and D. T. Matt. 2016. Sustainable production in emerging markets through distributed manufacturing systems (DMS). *Journal of Cleaner Production* 135:127–138.
43. Robertson, M. 2021. *Sustainability principles and practice[M]*. London: Routledge.
44. Jorgensen, S. E. 2009. *Ecosystem ecology[M]*. Academic Press.

45. Pryshlakivsky, J., and C. Searcy. 2013. Sustainable development as a wicked problem[M]. In *Managing and engineering in complex situations*, ed. S. F. Kovacic and A. Sousa-Poza, 109–128. Berlin: Springer.
46. Shedroff, N. 2019. *Design is the solution: The future of design will be sustainable [M/OL].* New York: Rosenfeld Media. https://nathan.com/design-is-the-solution/.
47. Simanis, E., S. Hart, and D. Duke. 2008. The base of the pyramid protocol: Beyond "basic needs" business strategies. *Innovations: Technology, Governance, Globalization* 3 (1): 57–84.
48. Beckert, S. 2019. *Empire of cotton: A new history of global capitalism[M].* Penguin.
49. Srai, J. S., M. Kumar, G. Graham, et al. 2016. Distributed manufacturing: Scope, challenges and opportunities. *International Journal of Production Research* 54 (23): 6917–6935.
50. Norman, D. 2003. *Emotional design (1st ed.) [M].* Basic Books.
51. Worster, D. 1994. *Nature's economic: a history of ecological ideas (2nd ed.)[M].* Cambridge University Press.
52. Tischner, U. 2010. *Case studies in sustainable consumption and production[M].* Sheffield, UK: Greenleaf Publishing Limited.
53. Trist, E. L. 1981. *The evolution of socio-technical systems[M].* Toronto: Ontario Quality of Working Life Centre.
54. Markussen, T., Z. X. Sun, and X. Y. Xin. 2015. The disruptive aesthetics of design activism: Enacting design between art and politics[J]. *Design Issues* 2:4–10.
55. Vezzoli, C., F. Ceschin, J. C. Diehl, et al. 2015. New design challenges to widely implement"sustainable product-service systems"[J]. *Journal of Cleaner Production* 97 (15): 1–12.
56. Vezzoli, C., C. Kohtala, A. Srinivasan, et al. 2017. *Product-service system design for sustainability[M].* London: Routledge.
57. Vezzoli, C., and E. Manzini. 2018. *Design for environmental sustainability[M].* Berlin: Springer.
58. Papanek, V. 2005. *The green imperative: Ecology and ethics in design and architecture[M].* Thames & Hudson.
59. Waizsacker, E., and A. Wijkman. 2018. *Come On!: Capitalism, Short-termism, Population and the Destruction of the Planet(1st ed.)[M].* Springer.
60. World Design Organization. Industrial design[EB/OL]. 2015 [2021-08-23]. https://wdo.org/glossary/industrial-design/.
61. Yu, M. C. 2004. *Environmental ethics[M].* Beijing: Higher Education Press.
62. Mcneill, J. R. 2016. *The great acceleration: An environmental history of the Anthropocene since 1945.* 1st ed. Belknap Press: An Imprint of Harvard University Press.
63. Thackara, J. 2016. *How to thrive in the next economy: designing tomorrow's world today (1st ed.) [M].* Thames & Hudson.
64. Zhong, F., and E. Manzini. 2021. Design for Social Innovation: A social systematic perspective[J]. *Zhuangshi* 12:40–46.
65. Zhu, D. J. 2013. Beyond growth: What are differences between sustainable development economics and neoclassical economics[J]. *Academic Monthly* 10:79–89.

The manufacturer's authorised representative in the EU is Springer Nature Customer Service Centre GmbH, Europaplatz 3, 69115 Heidelberg, Germany. If you have any concerns regarding our products, please contact ProductSafety@springernature.com

Printed and bound by CPI Group (UK) Ltd, Croydon, CR0 4YY
30/09/2026
02225296-0009